Manual for the of steelwork b structures to Eurocode 3
October 2010

The **Institution**
of **Structural**
Engineers

Constitution of the Task Group

J D Parsons BSc(Hons) CEng FIStructE MICE *Chairman*
Dr J B Davison BEng PhD CEng MICE
I G Hill BEng(Hons) CEng FIStructE MICE
Dr D Bruyere BEng MEng PhD CEng MICE
Dr R J Pope MA MSc DPhil CEng FIStructE FIMechE MCIArb
M P Stephens BEng CEng MIStructE
F E Weare MSc DIC DMS CEng FIStructE MICE MIM MIHT

Consultant
M J A Banfi MA CEng FIStructE

Secretary to the Task Group
B Chan BSc(Hons) AMIMechE

Published by The Institution of Structural Engineers
International HQ, 11 Upper Belgrave Street, London SW1X 8BH
Telephone: +44 (0)20 7235 4535 Fax: +44 (0)20 7235 4294
Email: mail@istructe.org Website: www.istructe.org
First published 2010
ISBN 978-1-906335-16-8

© 2010 The Institution of Structural Engineers

The Institution of Structural Engineers and the members who served on the Task Group which produced this Manual have endeavoured to ensure the accuracy of its contents. However, the guidance and recommendations given should always be reviewed by those using the Manual in the light of the facts of their particular case and any specialist advice. No liability for negligence or otherwise in relation to this Manual and its contents is accepted by the Institution, the members of the Task Group, its servants or agents. **Any person using this Manual should pay particular attention to the provisions of this Condition.**
No part of this publication may be reproduced, stored in a retrieval system or transmitted in any form by any means without prior permission of the Institution of Structural Engineers, who may be contacted at 11 Upper Belgrave Street, London SW1X 8BH.

Contents

Tables x

Notation xii
Latin upper case letters xii
Latin lower case letters xiv
Greek upper case letters xvi
Greek lower case letters xvi

Foreword xviii

1 **Introduction** 1
1.1 Aims of the *Manual* 1
1.2 Eurocode system 1
1.3 Scope of the *Manual* 4
1.4 Contents of the *Manual* 4
1.5 Terminology 5
    1.5.1 General 5
    1.5.2 Changes of axes nomenclature 6
    1.5.3 Slenderness 6
1.6 Non-contradictory complementary information (NCCI) 7

2 **General principles** 8
2.1 Designing for safety 8
2.2 Design process 9
2.3 Stability 10
    2.3.1 Multi-storey braced structures 10
    2.3.2 Single-storey structures 10
    2.3.3 Forms of bracing 11
2.4 Robustness 11
2.5 Movement joints 11
2.6 Loading 12
    2.6.1 Permanent and variable actions 12
    2.6.2 Equivalent forces due to imperfections 13
2.7 Second order effects 16
2.8 Limit states 17
    2.8.1 General 17
    2.8.2 Ultimate limit state 17
    2.8.3 Serviceability limit states 18

Contents

2.9 Material properties 21
    2.9.1 Partial factors for materials 21
    2.9.2 Design strength 21
    2.9.3 Other properties 22
    2.9.4 Properties in fire 22
    2.9.5 Brittle fracture 22
    2.9.6 Prevention of lamellar tearing 24
2.10 Section classification 26
2.11 Methods of analysis 27
    2.11.1 General approach 27
    2.11.2 Joint modelling 27

3 **Braced multi-storey buildings – General** 29
3.1 Introduction 29
3.2 Actions (loads) 29
3.3 Material selection 30
3.4 Structural form and framing 30
    3.4.1 Stability 30
    3.4.2 Framing 33
    3.4.3 Flooring 34
3.5 Fire resistance 36
3.6 Corrosion protection 37
3.7 Robustness 39
    3.7.1 Introduction 39
    3.7.2 Horizontal ties 40
    3.7.3 Vertical ties 40
3.8 Connections 41

4 **Beams** 42
4.1 Non-composite beams with full lateral restraint 42
    4.1.1 Introduction 42
    4.1.2 Design procedure 42
4.2 Non-composite I and H section beams without full lateral restraint 45
    4.2.1 Introduction 45
    4.2.2 Design procedure 45
    4.2.3 Calculation of relative slenderness 46
    4.2.4 Beams with intermediate restraint 50
4.3 Other sections without full lateral restraint 52
    4.3.1 General 52
    4.3.2 Rolled channel sections subject to bending about the major axis 52
    4.3.3 Welded I or H sections with equal flanges 53
    4.3.4 Welded doubly symmetrical box sections 53
    4.3.5 Rolled hollow sections 54
    4.3.6 Plates and flats 54
    4.3.7 Tee sections 54
    4.3.8 Rolled angles 55
    4.3.9 General case 56

Contents

4.4 Composite beams 57
    4.4.1 Introduction 57
    4.4.2 Construction case 57
    4.4.3 Properties of concrete and reinforcement 57
    4.4.4 Effective concrete width 58
    4.4.5 Vertical shear resistance 58
    4.4.6 Bending resistance 59
    4.4.7 Shear connectors 61
    4.4.8 Spacing and minimum amount of shear connection 64
    4.4.9 Transverse shear 66
    4.4.10 Serviceability and deflections 67

5 Columns in braced multi-storey buildings 69
5.1 Columns 69
5.2 Column selection 69
5.3 Columns in simple construction 70
5.4 Design procedure 71
5.5 Columns with additional moments 73

6 Members in bracing systems 75
6.1 Introduction 75
6.2 Cross sectional areas 75
6.3 Buckling lengths and slenderness ratio 76
6.4 Resistance of sections in axial compression 77
    6.4.1 General 77
    6.4.2 Angles in compression 77
6.5 Resistance of sections in axial tension 78
    6.5.1 General 78
    6.5.2 Angles connected by one leg 79
6.6 Bending and axial tension (EC3 Part 1-1 cl.6.2.9) 80

7 Single-storey buildings – General 81
7.1 Introduction 81
7.2 Actions (loads) 81
    7.2.1 General 81
    7.2.2 Permanent actions G 81
    7.2.3 Variable actions Q 82
    7.2.4 Equivalent horizontal forces 82
    7.2.5 Load combinations 82
7.3 Material selection 83
7.4 Structural form and framing 83
7.5 Fire resistance 84
7.6 Corrosion protection 84
7.7 Bracing 85
7.8 Purlins and side rails 86
7.9 Roof and cladding materials 86

Contents

8 Portal frames with pinned bases 88
8.1 Introduction 88
8.2 Plastic analysis 89
8.3 Sizing of rafters and columns 89
8.4 Design procedure 89
8.5 Sway and snap-through stability 92
    8.5.1 General 92
    8.5.2 Sway stability check 92
    8.5.3 Snap-through stability check 94
8.6 Serviceability check – deflection 94
8.7 Check on position of plastic hinge in rafter and calculation of load resistance 95
8.8 Stability checks 96
    8.8.1 Restraint of plastic hinges 96
    8.8.2 Rafter stability 98
    8.8.3 Stability of haunch 98
    8.8.4 Stability of column 100

9 Lattice girders and trusses with pin-base columns 101
9.1 Lattice girders and trusses 101
9.2 Determination of section sizes 104
9.3 Columns for single-storey buildings braced in both directions 105
9.4 Columns for single-storey buildings braced in one direction only in the side walls and/or in the valleys 106

10 Connections 107
10.1 Introduction 107
10.2 Connection design 108
10.3 Bolts 109
    10.3.1 Spacing and edge distances 109
    10.3.2 Dimensions of holes 110
    10.3.3 Design resistance of ordinary bolts 110
    10.3.4 Bolts in slip-resistant connections (Category B and C connections) 113
10.4 Welds 115
    10.4.1 Fillet welds 115
    10.4.2 Butt welds 116

11 Typical connection details 118
11.1 Introduction 118
11.2 Column bases 118
    11.2.1 General 118
    11.2.2 Design of base plates 119
    11.2.3 Design of bases (moment + axial load) 120

Contents

11.3 Beam-to-column and beam-to-beam connections in simple construction 121
    11.3.1 Beam-to-column web cleats 121
    11.3.2 Beam-to-beam web cleats 122
    11.3.3 Beam-to-column flexible end plates 123
    11.3.4 Beam-to-beam flexible end plates 123
    11.3.5 Fin plates 124
    11.3.6 Block tearing 124
11.4 Column-to-column splices 126
    11.4.1 Column bearing splices 127
    11.4.2 Column non-bearing splices 127
11.5 Portal frame eaves connections 129
    11.5.1 General 129
    11.5.2 Geometry – eaves haunch 130
    11.5.3 Resistance of column flange in bending 131
    11.5.4 Effective length of equivalent T-stub 132
    11.5.5 Web tension in beam or column 132
    11.5.6 Final resistance moment 132
    11.5.7 Compression checks 133
    11.5.8 Column web panel shear 134
    11.5.9 Vertical shear 134
    11.5.10 Weld sizes 135
    11.5.11 Summary of procedure 135
11.6 Resistance to transverse forces i.e web bearing and buckling 136

References 137

Appendix A Moment resistances of UB sections 141

Appendix B Resistances of UC sections 148

Tables

Table 2.1 Values of initial imperfection ϕ 13
Table 2.2 Partial load factors 17
Table 2.3 Typical loadcase combinations 18
Table 2.4 Vertical deflection limits 19
Table 2.5 Horizontal deflection limits 19
Table 2.6 Nominal strengths of steels 21
Table 2.7 Maximum thickness for steelwork in buildings 23
Table 2.8 Selection of width to thickness ratios for classification of elements 26
Table 3.1 Details of typical flooring systems and their relative merits 35
Table 3.2 Fire protection: type of protection 36
Table 3.3 Specifications from CIRIA 174 38
Table 3.4 Life to first maintenance for external specifications 38
Table 3.5 Environments and life to first maintenance for internal specifications 38
Table 3.6 Building classification for robustness 39
Table 4.1 Section classification for bending only 43
Table 4.2 Simplified expressions for relative slenderness 47
Table 4.3 Determination of C_1 48
Table 4.4 Values of slenderness parameter V 49
Table 4.5 Parameters k and D for simple beams without intermediate restraint 50
Table 4.6 Parameters k and D for a cantilever without immediate restraint 51
Table 4.7 Properties of normal weight concrete 58
Table 4.8 Properties of lightweight concrete 58
Table 4.9 Characteristic resistance of studs 62
Table 4.10 Percentage reinforcement 66
Table 5.1 UB Sections which are not class 4 in direct compression 70
Table 5.2 Selection of buckling curve 73
Table 5.3 Uniform moment factors 74
Table 6.1 Reduction factors β_2 and β_3 80
Table 7.1 Load combinations for single-storey buildings 82
Table 7.2 Typical spacings and spans for single-storey buildings 84
Table 7.3 Lightweight roofing systems and their relative merits 87
Table 10.1 Type of joint model 107
Table 10.2 Bolt spacing and end distance 110
Table 10.3 Nominal clearances for bolts (mm) 110
Table 10.4 Resistance of bolts 111
Table 10.5 Nominal yield strength and ultimate strengths for bolts (used as characteristic values in calculations) 112
Table 10.6 Shear and tensile resistance for class 8.8 bolts 112
Table 10.7 Bearing resistance of class 8.8 ordinary bolts: grade S275 material 113
Table 10.8 Bearing resistance of class 8.8 ordinary bolts: grade S355 material 113
Table 10.9 Resistance of pre-loaded bolts in friction connections 114

Tables

Table 10.10 Slip resistance of preloaded bolts   114
Table 10.11 Calculation of resistance of fillet welds   116
Table 10.12 Fillet weld resistances   116
Table 11.1   Concrete design values and bearing strengths   120
Table 11.2   Expressions for m and e 130
Table 11.3   Resistance of T-stub   131
Table 11.4   Maximum column flange thickness in mm for plastic distribution of bolt forces   133
Table A1   S275 Steel – 457 deep and above   141
Table A2   S275 Steel – 406 deep and below   143
Table A3   S355 Steel – 457 deep and above   144
Table A4   S355 Steel – 406 deep and below   146
Table B1   S355 Steel   148

Notation

Latin upper case letters

A	Cross-sectional area
A_b	Bearing area under baseplate
A_c	Cross-sectional area of concrete
A_{c0}	Concrete area subject to compression
A_d	Design value of accidental action
A_{eff}	Effective area of a class 4 cross section
A_{net}	Net area of section
A_{nt}	Net area of connected part subject to tension
A_{nv}	Net area of connected part subject to shear
A_s	Tensile stress area of a bolt
A_v	Shear area of a structural steel section
$B_{p,Rd}$	Punching shear resistance of a bolt
C	Compression force on concrete under a baseplate
C_1	Factor to allow for the shape of the moment diagram in the calculation of the elastic critical moment
C_{my}	Equivalent uniform moment factor
C_{mz}	Equivalent uniform moment factor
D	Destabilising parameter
E	Modulus of elasticity
E_{cm}	Short term secant modulus of elasticity of concrete
E_{lcm}	Short term secant modulus of elasticity of lightweight concrete
F	Design load on an element
F_0	Maximum uniform distributed load for plastic failure of a portal frame rafter as a fixed beam
$F_{p,Cd}$	Design preloading force
F_r	Factored vertical load on portal frame rafter
F_{Rdu}	Concentrated design resistance force
$F_{s,Rd}$	Slip resistance of a bolt
$F_{t,Ed}$	Tensile force on a bolt
$F_{t,Rd}$	Tensile resistance of a bolt
$F_{tr,Rd}$	Tensile resistance of a bolt row
$F_{T,Rd}$	Resistance of a t-stub
$F_{v,Ed}$	Shear force on a bolt
$F_{v,Rd}$	Shear resistance of a bolt
$F_{w,Rd}$	Shear resistance per unit length of a weld
G	Shear modulus
G_k	Characteristic value of permanent actions
H	Horizontal reaction on portal frame
H_{Ed}	Design value of the horizontal reaction at the bottom of the storey
H_{FR}	Horizontal force ratio for portal frame
I	Second moment of area
I_c	Second moment of area of column of portal frame

Notation

I_r	Second moment of area of rafter of portal frame at its shallowest point
I_t	St. Venant torsion constant
I_w	Warping constant
I_y	Second moment of area (Inertia) about the y-y axis
I_z	Second moment of area (Inertia) about the z-z axis
K_1	Factor for stability of portal frame haunch
L	Length; span; effective span
L_{eff}	Effective length; effective length of yield line in equivalent T-stub
L_h	Length of haunch
L_k	Minimum length between compression flange restraints for uniform sections
L_m	Length between restraints adjacent to plastic hinge
L_s	Minimum length between compression flange restraints for haunched sections
L_y	Length between compression flange restraints
M	Bending moment
M_a	Contribution of the structural steel section to the design plastic resistance moment of the composite section
$M_{b,Rd}$	Design value of the buckling resistance moment of a beam
M_{cr}	Elastic critical moment for lateral-torsional buckling of a beam
M_{Ed}	Design bending moment
M_h	Maximum end moment
M_p	Design value of the plastic resistance moment of equivalent T-stub taking into account the compressive normal force
M_{pl}	Moment ratio for column of a portal frame
$M_{pl,Rd}$	Design value of the plastic resistance moment of the composite section with full shear connection
M_{pr}	Moment ratio for rafter of a portal frame
M_s	Midspan moment
$M_{u,Ed}$	Design moment about u-u axis
$M_{v,Ed}$	Design moment about v-v axis
$M_{y,Ed}$	Design bending moment applied to the section about the y-y axis
$M_{y,Rd}$	Design value of the resistance moment of a section about the y-y axis
$M_{z,Ed}$	Design bending moment applied to the section about the z-z axis
N	Compressive normal force
$N_{b,Rd}$	Design buckling resistance
N_c	Design value of the compressive normal force in the concrete flange
$N_{c,Rd}$	Design compression resistance
N_{cr}	Elastic critical normal force
N_{Ed}	Design value of the axial force
$N_{pl,a}$	Design value of the plastic resistance of the structural steel section to normal force
$N_{pl,Rd}$	Design value of the plastic resistance of the composite section to compressive normal force
$N_{u,Rd}$	Ultimate resistance of net section
$P_{pb,Rd}$	Design value of the bearing resistance of a stud
P_{Rd}	Design value of the shear resistance of a single connector
P_{Rk}	Characteristic value of the shear resistance of a single connector
Q_k	Characteristic value of variable actions
R_c	Axial resistance of concrete in the slab of a composite beam
R_w	Resistance of weld connecting the end plate to beam web

Notation

S_k	Characteristic value of snow loads
T_1	Force required in internal ties
U	Section parameter
V	Slenderness parameter; Shear
V_{Ed}	Total design value vertical load on the structure at the bottom of the storey
$V_{eff,Rd}$	Design block tearing resistance
$V_{pl,Rd}$	Design value of the plastic shear resistance of the section
W	Section modulus
W_{el}	Elastic section modulus
W_k	Characteristic value of wind loads
W_{pl}	Plastic section modulus
$W_{pl,y,c}$	Plastic section modulus of column of portal frame
$W_{pl,y,r}$	Plastic section modulus of rafter of portal frame
$W_{u,min}$	Minimum elastic modulus for bending about the major principal axis

Latin lower case letters

a	Ratio of web area to gross area; distance between centroid of steel section and centre of concrete slab; throat thickness of weld
a_b	Factor for bearing resistance of bolts
a_d	Factor for bearing resistance of bolts
b	Width of a cross section; width of slab
b_{eff}	Total effective width
b_f	Width of the flange of a steel section
b_p	Width of end plate
b_r	Width of rib of profiled steel sheeting
b_0	Mean width of a concrete rib (minimum width for re-entrant sheeting profiles); width of haunch
c	Clear width of an internal element or outstand of a steel section; effective outstand of a baseplate, taper factor
d	Clear depth of the web of the structural steel section; diameter of the shank of a stud connector; overall diameter of circular hollow steel section; minimum transverse dimension of a column; diameter of bolt
d_0	Hole diameter
d_E	Deflection of portal frame
e	Eccentricity of loading; edge distance of bolts
e_s	Shrinkage strain
e_0	Maximum amplitude of a member imperfection
e_1	End distance for bolt i.e. in the direction of the applied load
e_2	Edge distance for a bolt i.e. perpendicular to the direction of the applied load
f_{cd}	Design value of the compressive strength of concrete
f_{ck}	Characteristic value of the cylinder compressive strength of concrete at 28 days
f_{cu}	Characteristic value of the cube compressive strength of concrete at 28 days
f_{jd}	Design bearing strength

Notation

f_{lck}	Characteristic value of the cylinder compressive strength of lightweight concrete at 28 days
f_{lcu}	Characteristic value of the cube compressive strength of lightweight concrete at 28 days
f_{sd}	Design value of the yield strength of reinforcing steel
f_{sk}	Characteristic value of the yield strength of reinforcing steel
f_u	Specified ultimate tensile strength
f_{ub}	Ultimate tensile strength of bolt
f_y	Nominal value of the yield strength of structural steel
f_{yd}	Design value of the yield strength of structural steel
$f_{y,wc}$	Yield strength of a column web
g	Factor on ratio of inertias of cross-section
g_k	Permanent distributed load on the floor
h	Depth of a cross section; storey height; height of the structure; overall depth; thickness
h_c	Minimum depth of rafter in a portal frame; thickness of concrete above the main flat surface of the top of the ribs of the sheeting; depth of column in portal frame
h_h	Additional depth of haunch in a portal frame
h_p	Overall depth of the profiled steel sheeting excluding embossments
h_s	Depth of portal frame rafter allowing for the slope
h_{sc}	Overall nominal height of a stud connector
i	Radius of gyration about the relevant axis
i_z	Radius of gyration about the z-z axis
k	Effective length parameter
k_1	Factor for bearing resistance of bolts
k_j	Concentration factor for concrete bearing strength
k_l	Reduction factor for resistance of a headed stud used with profiled steel sheeting parallel to the beam
k_t	Reduction factor for resistance of a headed stud used with profiled steel sheeting transverse to the beam
m	Number of columns in a row; shear span for end plate connection
n	Modular ratio; number of shear connectors; number of friction interfaces; distance used for calculation of resistance of T-stub in tension
n_r	Number of stud connectors in one rib
n_s	Number of shear connectors
p	Spacing of two bolt holes perpendicular to the axis of the section, bolt pitch
p_1	Bolt spacing in the direction of the applied load
p_2	Bolt spacing perpendicular to the direction of the applied load
q_{ed}	Equivalent design force per unit length
q_k	Variable distributed load on the floor
r	Rise of a portal frame
s	Longitudinal spacing centre-to-centre of the stud shear connectors; slip; leg length of fillet weld; length of fusion face; spacing of ties; staggered pitch of bolt holes
t	Thickness
t_f	Thickness of a flange of the structural steel section
t_p	Thickness of packing in a bolted connection
t_w	Thickness of the web of the structural steel section
t_{wb}	Thickness of the web of the structural steel beam

Notation

t_{wc}	Thickness of the web of the structural steel column
u-u	Major principal axis (where this does not coincide with the y-y axis)
v-v	Minor principal axis (where this does not coincide with the z-z axis)
v_a	Factor for calculation of slenderness of an unequal angle
w	Factor for calculation of slenderness of a tee section, gauge of bolts
w'	Total load on a portal frame at failure
x	Depth of concrete stress block; section property; distance from column to the point of maximum moment in a portal frame
x-x	Axis along a member
y-y	Axis of a cross-section
z-z	Axis of a cross-section

Greek upper case letters

Ω	Arching effect ratio for portal frame

Greek lower case letters

α	Coefficient of linear thermal expansion; Factor; parameter
α_{cc}	Coefficient used to calculate the design compressive strength of concrete
α_{cr}	Factor by which the design loads would have to be increased to cause elastic instability
α_h	Reduction factor on global imperfection for height h applicable to columns
α_m	Reduction factor on global imperfection for the number of columns in a row
β_w	Ratio of moment resistance to plastic moment resistance of section; weld correlation factor
γ_C	Partial factor for concrete
γ_G	Partial factor for permanent loads
γ_M	Partial factor for a material property, also accounting for model uncertainties and dimensional variations
γ_{M0}	Partial factor for structural steel applied to resistance of cross-sections, see BS EN 1993-1-1, 6.1(1)
γ_{M1}	Partial factor for structural steel applied to resistance of members to instability assessed by member checks, see BS EN 1993-1-1, 6.1(1)
γ_{M2}	Partial factor for structural steel applied to resistance of cross sections in tension to fracture, see BS EN 1993-1-1, 6.1(1); partial factor for structural steel applied to connection resistance see BS EN 1993-1-8, 2.2(2)
γ_{M3}	Partial factor for structural steel applied to slip resistance of connections at ULS, see BS EN 1993-1-8, 2.2(2)
$\gamma_{M3,ser}$	Partial factor for structural steel applied to slip resistance of connections at SLS, see BS EN 1993-1-8, 2.2(2)
γ_Q	Partial factor for variable loads

Notation

γ_S	Partial factor for reinforcing steel
γ_v	Partial factor for design shear resistance of a headed stud
δ	Central deflection
$\delta_{H,Ed}$	Horizontal displacement at the top of the storey, relative to the bottom of the storey
δ_q	In-plane deflection of a bracing system
ε	$\sqrt{235/f_y}$, where f_y is in N/mm^2
η	Degree of shear connection; coefficient
λ	Slenderness
$\bar{\lambda}$	Relative slenderness
$\bar{\lambda}_{eff}$	Effective relative slenderness of angles
$\bar{\lambda}_{LT}$	Relative slenderness for lateral-torsional buckling
λ_1	Slenderness value to determine the relative slenderness
λ_p	Load factor for portal frame
λ_v	Slenderness relative to the v-v axis
λ_z	Slenderness relative to the z-z axis
μ	Coefficient of friction
ν	Poisson's ratio
ϕ	Global initial sway imperfection
ϕ_a	Equivalent slenderness coefficient for rolled angles
ϕ_0	Basic value for global initial sway imperfection
χ	Reduction factor for flexural buckling
χ_{LT}	Reduction factor for lateral torsional buckling
χ_{min}	Minimum value of reduction factor for flexural buckling
ψ	Ratio of end moments; monosymmetry index
ψ_0	Characteristic combination factor
ψ_1	Frequent combination factor

Foreword

The Eurocode for the Design of Steel Structures (Eurocode 3) and the UK National Annexes (NA) setting out the Nationally Determined Parameters (NDPs) have been published. These documents, together with BS EN 1990: 2002: Basis of Structural Design and BS EN 1991: 2002: Eurocode 1: Actions on Structures, and their respective NAs, provide a suite of information for the design of most types of steel building structures in the UK. From 2010 the current National Standards (e.g. BS 5950) will be withdrawn and replaced by the Eurocodes.

This *Manual* is intended to provide guidance on the design of many common steel building frames and to show how the provisions of BS EN 1993-1-1: 2005, BS EN 1993-1-8: 2005 and BS EN 1993-1-10: 2005 can be selected for that purpose. Steel buildings frequently include composite floors and the *Manual* also covers some aspects of BS EN 1994-1-1: 2004. Certain limitations have been introduced to simplify the design process and to select from the comprehensive provisions of EC3 in order to aid the process of familiarisation. Much useful advice contained within the 2008 'grey' book *Manual for the design of steelwork building structures* has been retained and will therefore be familiar to many users of this new *Manual*. Chapter headings appear similar to the old 'grey' book but opportunity has been taken to rationalise and reorder various sections.

In preparing this publication the Consultant has benefited from access to the Eurocode design guides on steel and composite structures produced by Corus, the British Constructional Steelwork Association and by the Steel Construction Institute. To each of these organisations the Institution accords its thanks.

Foreword

I thank all of the members of the Task Group, and their organisations, who have given their time voluntarily. Special thanks are given to Mike Banfi, of Arup, who acted as Consultant to the Task Group, and to Berenice Chan, secretary to the Task Group, for guiding us all through the drafting process and ensuring a degree of harmony with the Manuals of other materials. During the review process, members of the Institution provided invaluable comment on the draft *Manual* that has contributed to its improvement.

I join with all of the other members of the Task Group in commending this *Manual* to the industry.

J. D. Parsons

James Parsons
Chairman

1 Introduction

1.1 Aims of the *Manual*

This *Manual* provides guidance on the design of single and multi-storey building structures using structural steelwork. Structures designed in accordance with this *Manual* should comply with BS EN 1993-1-1: 2005[1] (EC3 Part 1-1), BS EN 1993-1-8: 2005[2] (EC3 Part 1-8), BS EN 1993-1-10: 2005[3] (EC3 Part 1-10) and BS EN 1994-1-1: 2004[4] (EC4) where applicable. It assumes that the Nationally Determined Parameters from the UK National Annexes are applied. It includes the corrigenda issued by BSI to Eurocode 3 and 4 up to the end of February 2010.

It is primarily intended for those carrying out hand calculations, especially for initial layout and/or sizing of members. It is not necessarily relevant to computer analysis. However it is good practice that such hand analysis methods are used to verify the output of more sophisticated methods.

In many cases the expressions in the *Manual* are not those that appear in the Eurocodes. Where they are exactly the same as those in the Eurocodes they are referenced by an equation number and a reference to the Eurocode part e.g. (5.5 EC3-1-1) is a reference to equation 5.5 in Eurocode 3 Part 1-1[1].

1.2 Eurocode system

The structural Eurocodes were initiated by the European Commission but are now produced by the Comité Européen de Normalisation (CEN) which is the European standards organization, its members being the national standards bodies of the EU and EFTA countries, e.g. BSI.

The complete set of Eurocodes consists of the following:
BS EN 1990:   Eurocode: Basis of structural design (EC0)
BS EN 1991:   Eurocode 1: Actions on structures (EC1)
              Part 1-1:   General actions – densities, self-weight and imposed loads
              Part 1-2:   General actions – actions on structures exposed to fire
              Part 1-3:   General actions – snow loads
              Part 1-4:   General actions – wind loads

1.2 Introduction

                Part 1-5: General actions – thermal actions
                Part 1-6: General actions – actions during execution
                Part 1-7: General actions – accidental actions
                Part 2: Traffic loads on bridges
                Part 3: Actions induced by cranes and machinery
                Part 4: Actions on structures – silos and tanks
BS EN 1992: Eurocode 2: Design of concrete structures (EC2)
BS EN 1993: Eurocode 3: Design of steel structures (EC3)
                Part 1-1: General rules and rules for buildings
                Part 1-2: Structural fire design
                Part 1-3: Cold-formed thin gauge members and sheeting
                Part 1-4: Stainless steels
                Part 1-5: Plated structural elements
                Part 1-6: Strength and stability of shell structures
                Part 1-7: Strength and stability of planar plated structures transversely loaded
                Part 1-8: Design of joints
                Part 1-9: Fatigue strength of steel structures
                Part 1-10: Selection of steel for fracture toughness and through-thickness properties
                Part 1-11: Design of structures with tension components made of steel
                Part 1-12: Supplementary rules for high strength steel
                Part 2: Steel bridges
                Part 3: Towers, masts and chimneys
                Part 4: Silos, tanks and pipelines
                Part 5: Piling
                Part 6: Crane supporting structures
BS EN 1994: Eurocode 4: Design of composite steel and concrete structures (EC4)
                Part 1-1: General rules and rules for buildings
                Part 1-2: Structural fire design
                Part 2: General rules and rules for bridges
BS EN 1995: Eurocode 5: Design of timber structures (EC5)
BS EN 1996: Eurocode 6: Design of masonry structures (EC6)
BS EN 1997: Eurocode 7: Geotechnical design (EC7)
BS EN 1998: Eurocode 8: Design of structures for earthquake resistance (EC8)
BS EN 1999: Eurocode 9: Design of aluminium structures (EC9)

All the Eurocodes are in several parts but only EC1, EC3 and EC4 have been listed in full.

The European standards for design and materials relating to EC3 and EC4 are shown in Figure 1.1.

All Eurocodes follow a common editorial style. The codes contain 'Principles' and 'Application rules'. Principles are identified by a letter P following the paragraph number. Principles are general statements and definitions for

Introduction 1.2

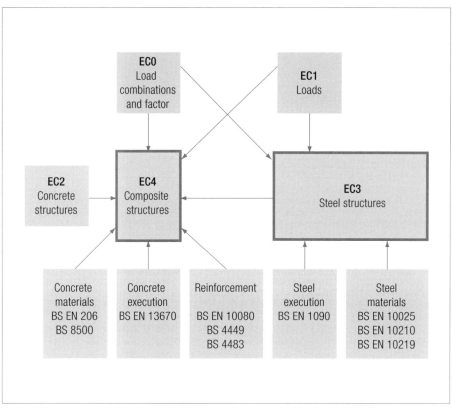

Fig 1.1 Standards relating to EC3 and EC4

which there is no alternative, as well as requirements and analytical models for which no alternative is permitted unless specifically stated.

Application rules are generally recognised rules which comply with the Principles and satisfy their requirements. Alternative rules may be used provided that compliance with the Principles can be demonstrated, however the resulting design cannot be claimed to be wholly in accordance with the Eurocode, though it will remain in accordance with the Principles.

Each Eurocode gives values with notes indicating where national choice may have to be made. These are recorded in the National Annex for each Member State as Nationally Determined Parameters (NDPs).

1.3 Scope of the *Manual*

The range of structures covered by the *Manual* are:
- braced low to medium rise multi-storey structures that do not rely on bending resistance of columns for their overall stability. Some guidance is given on allowance for second order effects for limited situations, other situations are outside the scope of this *Manual*
- single-storey structures using portal frames, posts and latticed trusses or posts and pitched roof trusses.

In using the *Manual* the following should be noted:
- The steel material is assumed to be either S275 or S355.
- Structures requiring seismic resistant design are not covered. Refer to BS EN 1998-1:2004[5] (EC8).
- Elements susceptible to fatigue are not covered.
- Normal structure/cladding interfaces and finishes are assumed.
- Apart from some guidance on floor vibrations, the dynamic performance of structures is not covered.
- Sections with class 4 (slender) elements are not covered.
- Sections where the shear buckling resistance of the web needs to be checked are not covered.
- Except for portal frames, elastic methods of global analysis are assumed.

In the design of structural steelwork it is not practical to include all the information necessary for section design within the covers of one book. Section properties and resistances have been included in the *Manual* where appropriate, but nevertheless reference will need to be made to *Steel building design: Design data* SCI Publication No 363[6].

EC3 Parts 2 to 6 also covers structures which are outside the scope of this *Manual* e.g. towers, masts, chimneys, pipelines, silos and bridges.

1.4 Contents of the *Manual*

The *Manual* covers the following:
- guidance on structural form, framing and bracing, including advice on the selection of floors, roofing and cladding systems, and advice on deflection, thermal expansion, fire and corrosion protection
- step-by-step procedures for designing the different types of structure and structural elements including verification of robustness and design of connections.

Introduction 1.5

1.5 Terminology

1.5.1 General

In order to rationalise the meaning of various technical terms for easy translation, some of the terms used in the past have been modified and given precise meanings in the Eurocodes. The following, which are adopted in this *Manual*, are of particular importance in the understanding of Eurocodes:

Accidental action	Action, force or event, usually of short duration, which is unlikely to occur with significant magnitude over the period of time under consideration during the design working life. This will generally be impact, fire or explosion.
Buckling length	The distance between the points of contraflexure in the fully buckled mode of a compression member or flange. This will normally be the system length multiplied by an appropriate factor. This is also called the effective length.
Direct action	A force (load) applied to a structure.
Execution	The act of constructing the works. For steel structures this includes both fabrication and erection.
Frame	An assembly of members capable of carrying actions.
Global analysis	Any analysis of all or part of a structure. This includes beams in simple frames as well as the complete analysis of rigid jointed structures. The analysis may be either elastic or plastic.
Indirect action	An imposed deformation; such as temperature effects or settlement.
Limit states	States beyond which the structure no longer satisfies the design performance requirements.
Permanent actions	Dead loads, such as self-weight of the structure or fittings, ancillaries and fixed equipment.
Resistance	The strength of a member in a particular mode of failure.
Serviceability limit states	Correspond to limit states beyond which specified service criteria are no longer met, with no increase in action (load).
Variable action	Imposed loads, wind loads or snow loads.
Ultimate limit states	Those associated with collapse, or with other forms of structural failure that may endanger the safety of people.

The Institution of Structural Engineers Manual for the design of steelwork building structures to Eurocode 3

1.5 Introduction

1.5.2 Changes of axes nomenclature

The Eurocodes use a different axis terminology to that traditionally used in the UK, it matches the sign convention usually adopted by computer software. Figure 1.2 gives the notation used throughout the Eurocodes for member axes. The x-x axis lies along the length of the member. For sections like the I section shown in Figure 1.2 the y-y axis is the major axis and the z-z axis is the minor axis. Where the principal axes are not the same as the geometric axes of the section e.g. for the angle shown in Figure 1.2, the u-u axis is the major axis and v-v axis is the minor axis.

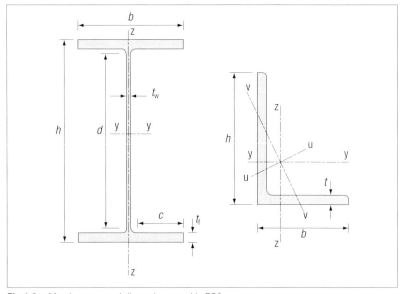

Fig 1.2 Member axes and dimensions used in EC3

1.5.3 Slenderness

The Eurocode uses a slenderness that is defined as the ratio of resistances rather than geometric values. It is called the relative or non-dimensional slenderness and is represented by the symbol $\bar{\lambda}$. The general relationship is:

$\bar{\lambda} = ((\text{Resistance without buckling})/(\text{Elastic critical buckling load}))^{0.5}$

Introduction 1.6

This *Manual* considers flexural and lateral torsional buckling and the expressions for each mode are:

$$\bar{\lambda} = \sqrt{\frac{Af_y}{N_{cr}}} \quad \text{for flexural buckling}$$

$$\bar{\lambda}_{LT} = \sqrt{\frac{Wf_y}{M_{cr}}} \quad \text{for lateral torsional buckling}$$

The values of *A* and *W* depend on the classification of the section (see Section 2.10). The area *A* used for class 1, 2 and 3 sections is the gross area. For class 4 sections (which are not covered by this *Manual*) an effective area is used to allow for local buckling. The modulus *W* is the plastic modulus for class 1 and 2 sections and the elastic modulus for class 3 sections. N_{cr} is the Euler buckling load for the member in compression and M_{cr} is the elastic critical buckling moment (see Section 4).

The relative slenderness and the geometrical slenderness are related by the expression:

$$\bar{\lambda} = \lambda \sqrt{\frac{f_y}{\pi^2 E}} \quad \text{and} \quad \bar{\lambda}_{LT} = \lambda_{LT} \sqrt{\frac{f_y}{\pi^2 E}}$$

1.6 Non-contradictory complementary information (NCCI)

The Eurocodes allow the National Annex to reference Non-Contradictory Complementary Information. For certain parts of the Eurocode, Published Documents have been produced and these are identified as NCCI in the appropriate National Annex. EC3 Part 1-10[3] is an example where this has happened (see reference 26). EC3 Parts 1-1[1] and 1-8[2] do not reference particular documents but give a reference to a website www.steel-ncci.co.uk and warn that the material on the site has not all been reviewed, it is not endorsed by the UK National committee and some may contain elements that conflict with the Eurocode. The material on the website may be helpful but the user should satisfy themselves of its fitness for their particular purpose.

A design that uses additional guidance can still comply with the Eurocodes provided the additional information is NCCI i.e. does not contradict the Eurocodes. The user must also satisfy themselves that any additional guidance is from an authoritative source.

2 General principles

2.1 Designing for safety

Whether enshrined in law, statutory obligations, contractual requirements or good practice, designing for the safe construction, use, maintenance and eventual demolition of the structure must be taken into account.

Important principles which should be implemented when carrying out the design are:
- Identification of design responsibilities within the construction chain from client to contractor to user.
- Hazard elimination and risk reduction (The UK Health and Safety Executive promotes the phrase 'Eliminate, Reduce, Inform, Control' to describe the order in which to address hazards and risks).
- Effective communication of salient information including design responsibilities, particularly at contractual interfaces.

These principles are exemplified in the UK in the Construction (Design and Management) Regulations 2007[7], which are supported by central[8] and industry[9] guidance.

The designer who is responsible for the overall stability of the structure should be clearly indentified. This designer should ensure the compatibility of the structural design and detailing between all those structural parts and components that are required for overall stability, even if some or all of the structural design and detailing of those structural parts and components is carried out by another designer. In particular, the engineer responsible for the design of the frame should ensure that the connection details reflect their design assumptions, including situations where more than one structural material is employed.

The BCSA document *Allocation of design responsibilities in constructional steelwork*[10] identifies the information required to be provided at two stages of the design process. Design stage 1 is information required to enable the steelwork to be costed and stage 2 is that required for construction.

2.2 Design process

At the beginning of the design the aim should be to establish a simple structural scheme that is practicable, sensibly economic and not unduly sensitive to the various changes that are likely to be imposed as the overall design develops. Loads should be carried to the foundation by the shortest and most direct routes. In constructional terms simplicity implies (among other matters) repetition, avoidance of congested, awkward or structurally sensitive details, with straightforward temporary works and minimal requirements for unorthodox sequencing to achieve the intended behaviour of the completed structure. Sizing of structural members should be based on the longest relevant spans (slabs and beams) and largest areas of roof and/or floors carried (beams, columns, walls and foundations). The same sections should be assumed for similar but less onerous cases as this saves design and costing time and is of actual advantage in producing visual and constructional repetition and hence, ultimately, cost benefits. Simple structural schemes are quick to design and easy to build. They may be complicated later by other members of the design team trying to achieve their optimum conditions, but a simple scheme provides a good 'benchmark'.

Scheme drawings should be prepared for discussion and budgeting purposes incorporating such items as:
– general arrangement of the structure
– bracing
– type of floor construction
– critical and typical beam and column sizes
– typical edge details
– critical and unusual connection details.

A description of the main elements of the structure should also be provided and this should include:
– vertical load system
– stability system
– provision of movement joints
– loading assumed including allowance for finishes, imposed loads and wind loads
– environment assumed and hence any corrosion protection system
– fire resistance periods and any assumption about method of protection
– deflection limits, both horizontally and vertically including elements supporting cladding
– vibration response
– material grades
– design codes and any particular guidance used
– foundations
– erection methods assumed.

2.3 General principles

This description should form part of the information to be provided to the CDM Co-ordinator for inclusion in the Health and Safety file.

When the comments of the other members of the design team have been received and assimilated, the main elements of the structure should be reviewed to see whether they are still appropriate and revised if necessary. When these main elements have been agreed the structural members can be checked and redesigned as necessary.

2.3 Stability

2.3.1 Multi-storey braced structures

Lateral stability in two directions approximately at right-angles to each other should be provided by a system of vertical bracing. Bracing can generally be provided in the walls enclosing the stairs, lifts, service ducts, etc. Additional bracing may also be provided within other internal or external walls. The vertical bracing should, preferably, be distributed throughout the structure so that the combined shear centre is located approximately on the line of the resultant on plan of the applied overturning forces. Where this is not possible, torsional moments may result which must be considered when calculating the load carried by each vertical bracing system. If the vertical bracing is relatively close to the shear centre of the building it may be sensitive to asymmetric load and appropriate loadcases will need to be considered. Vertical bracing should be effective throughout the full height of the building and connected to appropriately designed foundations. If it is essential for the bracing to be discontinuous at one level, provision must be made to transfer the forces to other braced bays. Bracing should be arranged so that the angle with the horizontal is not greater than 60° nor less than 30°.

The elements of vertical bracing should be linked together by horizontal bracing. This is usually achieved via the floor diaphragm, but if that is not adequate, plan bracing may be necessary. In any case the load paths for the forces applied to the vertical bracing from the floors should be checked.

2.3.2 Single-storey structures

Lateral stability to these structures should be provided in two directions approximately at right-angles to each other. This may be achieved by:
– rigid framing, or
– vertically braced bays in conjunction with plan bracing.

General principles 2.5

2.3.3 Forms of bracing

Bracing may be of the following forms:
- horizontal bracing
  - triangulated steel framing
  - concrete floors or roofs
  - adequately designed and fixed profiled steel decking.
- vertical bracing
  - triangulated steel framing
  - reinforced concrete walls preferably not less than 180mm in thickness
  - masonry walls preferably not less than 150mm in thickness adequately connected to the steel frames.

Walls should not be used as a principal means of vertical bracing if they can be removed at a later stage. It should be noted that temporary bracing may need to be provided during erection if permanent elements, on which stability relies, are not yet present or effective.

2.4 Robustness

The layout of the building should constitute a robust and stable structure under normal loading to ensure that in the event of misuse or accident, damage will not be disproportionate to the cause. All members of a structure should be effectively tied together in the longitudinal, transverse and vertical directions as set out in Section 3.7. In framing the structure care should be taken to avoid members whose failure would cause disproportionate collapse. Where this is not possible, alternative load paths should be identified or the member in question designed as a key element (see Section 3.7).

2.5 Movement joints

Joints should be provided to minimize the effects of movements arising from temperature variations and settlement. The effectiveness of movement joints depends on their location, which should divide the whole structure into a number of individual sections. The joints should pass through the whole structure above ground level in one plane. The structure should be framed on each side of the joint, and each section should be structurally independent and be designed to be stable and robust without relying on the stability of adjacent sections.

2.6 General principles

Joints may also be required where there is a significant change in the type of foundation, plan configuration or height of the structure. Where detailed calculations are not made in the design, joints to permit horizontal movement of 15 to 25mm should normally be provided in the UK at approximately 50m centres both longitudinally and transversely. For single-storey sheeted buildings it may be acceptable to increase the spacing up to 100m with bracing centrally located between movement joints. More detailed advice is available in Access Steel document SS017a[11]. It is necessary to incorporate joints in finishes and in the cladding at the movement joint locations, in addition to joints required by the type of cladding.

A gap should generally be allowed between steelwork and masonry cladding to allow for the movement of columns under loading (except where the panel acts as an infill shear wall).

2.6 Loading

2.6.1 Permanent and variable actions

The loads to be used in calculations are:
- Characteristic permanent action (dead load), G_k: the weight of the structure complete with finishes, fixtures and fixed partitions.
- The characteristic variable actions (live loads) Q_{ki} e.g. imposed floor loads, wind loads and/or snow loads. Where more than one variable action occurs simultaneously a leading variable action Q_{k1} is chosen and the other variable actions are reduced by an appropriate combination factor. Where it is not obvious which should be the leading variable action, each action should be checked in turn and the worst case taken.

For typical buildings these loads are found in parts of BS EN 1991: Eurocode 1 : Actions on structures (EC1):
- Part 1-1: General actions – Densities, self-weight and imposed loads[12]
- Part 1-3: General actions – Snow loads[13]
- Part 1-4: General actions – Wind loads[14].

Loads from soil are calculated using BS EN 1997: Eurocode 7: Geotechnical design (EC7)[15].

As mentioned in Section 2.3.1, if the bracing is close to the shear centre it may be sensitive to asymmetric loads. BS EN 1991-1-4[14] gives guidance on asymmetric loading to be considered. In many cases applying the wind load at an eccentricity of 10% of the building width will be an acceptable alternative.

General principles 2.6

2.6.2 Equivalent forces due to imperfections

2.6.2.1 Introduction
Allowance must be made for imperfections. These include residual stresses as well as geometrical imperfections. Global imperfections for frames and bracing systems must be taken into account in the analysis and this is usually done by converting the imperfections to equivalent horizontal forces.

2.6.2.2 Frame imperfections
The global imperfection for a frame is defined as a sway imperfection and an inclination of the frame given as:

$$\phi = \phi_o \, \alpha_h \, \alpha_m \qquad \text{(5.5 EC3-1-1)}$$

where:
ϕ_o is the basic value: $\phi_o = 1/200$
α_h is the reduction factor for height h applicable to columns:

$\alpha_h = 2/\sqrt{h}$ but $2/3 \leqslant \alpha_h \leqslant 1.0$

h is the height of the structure in metres
α_m is the reduction factor for the number of columns in a row:

$\alpha_m = \sqrt{0.5(1 + 1/m)}$

m is the number of columns contributing to the horizontal force on the bracing system including only those columns which carry a vertical load not less than 50% of the average value of the columns considered.

The value of ϕ for various combinations of the number of columns and building height is shown in Table 2.1.

Table 2.1 Values of initial imperfection ϕ

Building height (metres)	Number of columns m in the row			
	1	2	4	10
$\geqslant 9$	1/300	1/346	1/379	1/405
7	1/265	1/306	1/335	1/357
4	1/200	1/231	1/253	1/270
Note These values are derived from equation 5.5 in EC3 Part 1-1.				

Because at splices in columns a restraining force equal to $\alpha_m N_{Ed}/100$ must be considered, for equal storey heights there will be a force of $\alpha_m N_{Ed}/200$ in the bracing above and below the splice. With non-uniform storey heights the force in one storey can be larger. This restraining force is not additional to the equivalent horizontal forces from the frame imperfection described above. It is therefore recommended that at least for initial design the frame imperfection is taken as 1/200. This will be conservative for forces in the

2.6 General principles

columns and reduced values should be considered if these become critical e.g. uplift at bases.

When, as described in Section 2.3.1 the bracing is sensitive to asymmetric load, it may be necessary to consider the imperfections for different parts of the building to be in opposite directions.

The connection of each column to the floor diaphragm will need to provide sufficient restraint, even if there is no splice. EC3 Part 1-1[1] gives an initial equivalent force of 0.5% of the axial force in the column. The final force will depend on the stiffness of the restraint system and it is recommended that the connection should be checked for a force equal to 1% of the axial force in the column.

The global sway imperfection is converted into an equivalent horizontal force applied at each floor by multiplying the design loads by ϕ, see Figure 2.1.

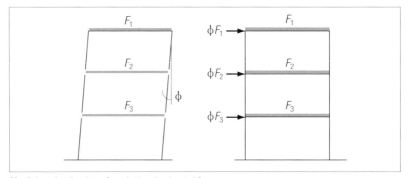

Fig 2.1 Application of equivalent horizontal forces

It should be noted that the equivalent horizontal forces are used in the frame analysis only, the resulting forces must be added to the horizontal loads when designing the frame for lateral loading. Although there is no overall force on the foundations there will be reactions for individual elements e.g. braced bays.

If the horizontal applied forces are very large the effect of imperfections will be small and EC3 Part 1-1[1] allows them to be neglected where the horizontal applied load is not less than 15% of the vertical load.

2.6.2.3 Imperfections for bracing

For the analysis of a bracing system that is required to provide lateral stability within the length of a beam or compression member, an initial bow imperfection e_0 should be used where $e_0 = \alpha_m L/500$. L is the span of the bracing system and α_m is as defined in Section 2.6.2.2 based on m being the number of elements to be restrained.

General principles 2.6

The bow imperfection can be replaced by an equivalent force q_{Ed} as shown in Figure 2.2:

$$q_{Ed} = \sum N_{Ed} 8(e_0 + \delta_q)/L^2$$

The deflection of the bracing, δ_q is that due to any applied loads plus the equivalent force. Provided δ_q is less than $L/2500$ then q_{Ed} can be taken as $\alpha_m \Sigma N_{Ed}/(50L)$. This can be used as an initial value and adjusted if necessary to suit the deflection calculated.

As mentioned in Section 2.6.2.2, at splices in the compression members a restraining force of $\alpha_m N_{Ed}/100$ must be considered. This is a local force applied to the bracing system and is not combined with the forces from the bow imperfection described above.

Local bow imperfections for individual members do not need to be taken into account in the global analysis except for members in moment frames subject to significant axial force. They do not need to be considered for braced multi-storey buildings described in Section 3. Further guidance is given in Part 1-1 of EC3[1].

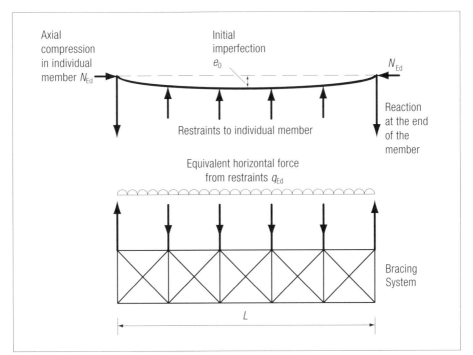

Fig 2.2 Equivalent force on a bracing system

2.7 Second order effects

Even in low rise buildings the additional forces due to the loads acting on the displaced structure may need to be considered. Where the structure is regular i.e. the distribution of horizontal load, vertical load and stability elements is the same in all storeys, simple methods may be used. Other cases are outside the scope of this *Manual*.

To establish the sensitivity of the frame to second order effects, the elastic critical load factor α_{cr} should be calculated. To do this the results of a linear elastic analysis can be used. The critical load factor can then be calculated for each storey using the expression:

$$\alpha_{cr} = \left(\frac{H_{Ed}}{V_{Ed}}\right)\left(\frac{h}{\delta_{H,Ed}}\right) \quad \text{(5.2 EC3-1-1)}$$

H_{Ed} and V_{Ed} equal the horizontal and vertical load at the base of the storey in the particular combination being considered, h is the storey height and $\delta_{H,Ed}$ is the horizontal deflection of the top of the storey relative to the bottom. The loads and hence the value of α_{cr} vary with load combination. The minimum value of α_{cr} should be taken as applying to the whole building.

If $\alpha_{cr} \geqslant 10$ then no account needs to be taken of second order effects. If $\alpha_{cr} < 3$ then second order analysis must be carried out. Otherwise account for second order effects can be made by multiplying the sway effects by $\alpha_{cr}/(\alpha_{cr}-1)$.

If the structure is symmetrical and subject to symmetrical vertical load it will not deflect horizontally under vertical loads and the sway effects are solely due to applied horizontal loads (including the equivalent horizontal loads calculated in Section 2.6.2). If the structure does deflect horizontally under vertical load the forces in the elements can be calculated by the following process:
(1) Carry out an analysis with all applied loads and horizontal supports applied at each storey
(2) Note the reactions at these horizontal supports
(3) Carry out an analysis with the actual loads and supports together with horizontal forces from the reactions calculated above. These horizontal forces should be multiplied by the factor $1/(\alpha_{cr}-1)$. This analysis gives the design forces.

2.8 Limit states

2.8.1 General

This *Manual* uses the limit state principle and the partial factor format common to all Eurocodes and as defined in BS EN 1990:2002[16] (EC0). Guidance on this document is included in the Institution Manual to EC1[17].

2.8.2 Ultimate limit state

The design loads are obtained by multiplying the characteristic loads by the appropriate partial factor from Table 2.2.

Table 2.2   Partial load factors

Load factor	Unfavourable value	Favourable value
Permanent actions γ_G	1.35	1.0
Variable actions γ_Q	1.50	0.0
Note Where equilibrium of the structure relies on self weight this must be considered using different load factors for permanent actions i.e. 1.1 for unfavourable and 0.9 for favourable actions.		

When more than one live load (variable action) is present the secondary live load may be reduced by the application of a combination factor ψ_0 (see below).

When considering structural effects EC0[16] gives two different sets of load combinations. This *Manual* uses the simpler expression (Eqn 6.10 in Annex A of EC0). The alternative expressions (Eqns 6.10a and 6.10b) may allow for more economy but are more complex.

From the expression in EC0[16], the load combination for a typical building becomes:

$$\gamma_G G_k + \gamma_Q Q_{k1} + \sum \gamma_Q \psi_0 Q_{ki}$$

where G_k is the action due to permanent load and Q_{k1}, Q_{k2} and Q_{k3} etc. are the actions due to vertical imposed loads, wind loads and snow etc., Q_{k1} being the leading action for the situation considered.

The 'unfavourable' and 'favourable' factors should be used so as to produce the most onerous condition. Generally permanent actions from a single load source may be multiplied by either the 'unfavourable' or the 'favourable' factor. For example all actions originating from the self weight of the structure may be considered as coming from one source and generally there is no requirement to consider different factors on different spans. Exceptions to this are where overall equilibrium is being checked and/or the structure is very sensitive to variations in permanent loads. See EC0[16].

2.8 General principles

The combination factor ψ_0 usually has a value of 0.7. Exceptions are:
- storage areas, $\psi_0=1.0$
- snow (buildings less than 1000m above sea level), $\psi_0=0.5$
- wind loads, $\psi_0=0.5$
- temperature effects (non fire), $\psi_0=0.6$

Generally, unless variable actions cannot occur together they must be considered in the combination. On roofs, where there may be loads due to snow, wind and imposed load, EC1 Part 1-1[12] says that imposed loads do not need to be applied together with snow or wind loads.

For a building subject only to imposed loads (not storage) and wind loads the loadcases that need to be considered are given in Table 2.3. The critical combination for an element will depend on the proportion of force from different actions e.g. for beams the combination with the imposed load as leading action is likely to be critical but for bracing elements the critical combination is likely to have wind as the leading action.

The Equivalent Horizontal Forces (EHF) appropriate to the vertical load need to be applied in all combinations.

Where the structure is subject to dancing, jumping or other dynamic loads during building use, the possibility of enhanced loads needs to be considered. Guidance is given in the National Annex to EC1 Part 1-1[12].

Table 2.3 Typical loadcase combinations

Leading variable action	Combination
Imposed load	$1.35G_k+1.5Q_k+0.75W_k$ (+EHF)[d]
Wind load	$1.35G_k+1.05Q_k+1.5W_k$ (+EHF)

Notes
a G_k is the characteristic permanent load.
b Q_k is the characteristic imposed load.
c W_k is the characteristic wind load.
d If the wind load has a favourable effect, a combination without wind will need to be considered.

2.8.3 Serviceability limit states

2.8.3.1 Deflection

The deflections and vibration of a structure should be limited so that deflections do not adversely affect the appearance or use of the structure, or cause damage to the finishes or non-structural elements.

The UK National Annex to EC3 Part 1-1[1] gives suggested limits on deflections under the characteristic (unfactored) variable loads i.e. excluding permanent loads. These are given in Tables 2.4 and 2.5.

General principles 2.8

Table 2.4   Vertical deflection limits

Member	Limit
Cantilevers	Length/180
Beams carrying plaster or other brittle finish	Span/360
Other beams	Span/200
Purlins and sheeting rails	To suit the characteristics of the particular cladding

Table 2.5   Horizontal deflection limits

Location	Limit
Tops of columns in single-storey buildings except portal frames	Height/300
Columns in portal frame buildings, not supporting crane runways	To suit the characteristics of the particular cladding
In each storey of a building with more than one storey	Height of that storey/300

Where two or more variable actions are considered (e.g. imposed floor load and wind) the secondary variable action(s) are multiplied by the ψ_0 combination factor. If it is not clear what is the leading variable action, multiple combinations should be considered.

It should be pointed out that the deflection criteria provided in any code can be used only as a guide to the serviceability of the structure and may not be taken as an absolute guide to satisfactory performance in all cases. It is the responsibility of the engineer to ensure that the limits used in the design are appropriate for the structure under consideration. It is common to limit deflections of beams supporting cladding to simplify the joints in the cladding system, absolute limits of 10mm have been used. This can be critical for edge beams.

2.8.3.2 Vibration
Vibration can occur in buildings due to:
– machinery
– wind
– movement of the occupants e.g. footfall
– being adjacent to highways and railways.

In cases where there is an induced oscillation arising from vibrating equipment, this should be controlled by ensuring that the frequency of the disturbing motion is either isolated or is not close to that of the structure or one of its harmonics. This may involve some complex analysis to derive the values of the frequencies; it should be noted that altering the safety factors will not guarantee a solution to the problem.

2.8 General principles

Guidance on the evaluation of the effects of vibration due to footfall is available in publications by the Concrete Centre[18] and SCI[19]. The effects will depend on:
- the excitation i.e. the source amplitude and frequency
- the mass of the floor
- the layout of the structure
- damping
- the stiffness (natural frequency) of the floor.

Office floors with low natural frequencies (e.g. below 3Hz) are unlikely to have adequate performance but providing beams with a higher frequency (e.g. 4 or 5Hz), will not guarantee there are no problems. Although typical large areas of floors for commercial buildings usually exhibit reasonable performance, additional care may be required for particular situations. Examples are premium space, heavily trafficked areas and small isolated bays. The criteria for residential buildings are more onerous than typical office floors.

Specific criteria are specified for hospital buildings, including very strict limits for operating theatres (HTM 08-01[20]).

Where there are rhythmic activities (e.g. dancing or co-ordinated exercise) the evaluation procedures are different. As well as addressing the response of the area of the activity the effect of the activity on adjacent areas that will be more sensitive must be considered.

Car park structures have traditionally been designed for a limit of 3Hz. Users are more tolerant of vibration in car parks than in offices, and structures designed for the traditional guidelines have performed adequately. Vibration from cars running over uneven surfaces or discontinuities should be avoided by detailing and workmanship.

Grandstands can be subject to crowd induced vibration and references should be made to the Institution's publication[21].

In connections subject to vibration, methods such as preloading, locking devices or welds, should be used to prevent the connection becoming loose. This is not necessary for typical floors with acceptable levels of vibration.

2.9 Material properties

2.9.1 Partial factors for materials

There are several partial material factors γ_M used in the Eurocode. The resistance of members and cross sections covered in EC3 Part 1-1[1] uses γ_{M0}, γ_{M1} and γ_{M2}. The factor γ_{M0} covers the general case while γ_{M1} covers the design of members where buckling takes place and γ_{M2} covers the case where the ultimate tensile strength of the material must be considered. According to the UK National Annex to EC3 Part 1-1, γ_{M0} and γ_{M1} are 1.0 and γ_{M2} is 1.1. Other γ values are used for the various facets of connection design and are defined in EC3 Part 1-8[2]. In particular the symbol γ_{M2} is also used in Part 1-8 and the UK National Annex to that part has chosen a value of 1.25 (see Section 10).

2.9.2 Design strength

According to the UK National Annex to EC3 Part 1-1[1] the strengths of materials to be used with the code are to be taken from the product standard. This is generally BS EN 10025[22] for open sections and plates, BS EN 10210[23] for hot finished hollow sections and BS EN 10219[24] for cold formed hollow sections. As well as giving the yield and tensile strength for the steel grades (e.g. S275 and S355) the material standards also give the impact strength for the subgrades (e.g. JR, J0 and J2). The strengths for various thicknesses are given in Table 2.6.

Table 2.6 Nominal strengths of steels

Steel grade	Thickness less than or equal to (mm)	Yield strength f_y (N/mm^2)	Ultimate tensile strength f_u (N/mm^2)
S275	16	275	410
	40	265	410
	63	255	410
	80	245	410
	100	235	410
	150	225	400
S355	16	355	470
	40	345	470
	63	335	470
	80	325	470
	100	315	470
	150	295	450

2.9 General principles

2.9.3 Other properties

Other properties of steel to be used in calculations are:
- Modulus of Elasticity $E = 210\,000$ N/mm^2
- Poisson's ratio in the elastic range $\nu = 0.3$
- Shear modulus $G = E/(2(1+\nu)) \approx 81\,000$ N/mm^2
- Coefficient of linear thermal expansion $\alpha = 12 \times 10^{-6}$ per K (for $T \leqslant 100°C$)
Note that for calculating the structural effects of unequal temperatures in composite concrete-steel structures to EC4[4] the coefficient of linear thermal expansion is taken as $\alpha = 10 \times 10^{-6}$ per K.

2.9.4 Properties in fire

The values above are for ambient temperature. The properties of steel change at elevated temperatures and the values for use in fire design are given in BS EN 1993-1-2:2005[25] (EC3 Part 1-2).

2.9.5 Brittle fracture

At low temperatures steel can become brittle and be susceptible to fracture. The risk of brittle fracture increases with the thickness of material and the level of tension stress, detail type and the rate of stress.

EC3 Part 1-10[3] deals with the choice of steel quality grade to avoid brittle fracture. The UK National Annex to that part has implemented changes to the safety allowance to reflect current UK practice. Table 2.7 is based on values in PD 6695-1-10[26] and gives maximum thicknesses of material for various grades in certain situations. The table is used as follows:
(1) In the top part of the table select a row corresponding to the detail type.
(2) Read across that row until you find a column with the appropriate ratio of applied tensile stress to yield.
(3) Look down that column into the lower part of the table until you find the row corresponding to the minimum temperature, the grade and the subgrade.
(4) The value gives the maximum thickness of material that can be used.

The table includes the two values of minimum temperature which are accepted in the UK as being suitable for internal and external situations. The detail type 'Plain material' applies to unwelded as-rolled, ground or machined surfaces. 'Bolted' applies to unwelded mechanically fastened joints or flame cut edges.

General principles 2.9

Table 2.7 Maximum thickness for steelwork in buildings

Detail type			Tensile stress level $\sigma_{Ed}/f_y(t)$								
Description		ΔT_{RD}									
Plain material		+30°C	≤0	0.15	0.30	≥0.5					
Bolted		+20°C		≤0	0.15	0.30	≥0.5				
Welded – moderate		0°C			≤0	0.15	0.30	≥0.5			
Welded – severe		-20°C				≤0	0.15	0.30	≥0.5		
Welded – very severe		-30°C					≤0	0.15	0.30	≥0.5	
Minimum temperature	Steel grade	Subgrade	Maximum thickness (mm)								
5°C (Internal)	S275	JR	122	102	85	70	60	50	40	32	27	22
		J0	192	172	147	122	102	85	70	60	50	40
		J2	200	200	192	172	147	122	102	85	70	60
		M, N	200	200	200	192	172	147	122	102	85	70
		ML, NL	200	200	200	200	200	192	172	147	122	102
	S355	JR	82	67	55	45	37	30	22	17	15	12
		J0	142	120	100	82	67	55	45	37	30	22
		J2	190	167	142	120	100	82	67	55	45	37
		K2,M,N	200	190	167	142	120	100	82	37	55	45
		ML, NL	200	200	200	190	167	142	120	100	82	67
-15°C (External)	S275	JR	70	60	50	40	32	27	22	17	12	10
		J0	172	147	122	102	85	70	60	50	40	32
		J2	200	192	172	147	122	102	85	70	60	50
		M, N	200	200	192	172	147	122	102	85	70	60
		MLNL	200	200	200	200	192	172	147	122	102	85
	S355	JR	45	37	30	22	17	15	12	10	7.5	5
		J0	120	100	82	67	55	45	37	30	22	17
		J2	167	142	120	100	82	67	55	45	37	30
		K2,M,N	190	167	142	120	100	82	67	55	45	37
		ML, NL	200	200	190	167	142	120	100	82	67	55

Note
This table is based on the following conditions:
(i) $\Delta T_{Rg} = 0$. See NA.2.1.1.3 of the National Annex to EC3 Part 1-10[3]. For further guidance on types of geometry where there is likely to be a stress concentration factor greater than unity see BS PD 6695-1-9[72], Figure 2b and c.
(ii) $\Delta T_r = 0$. See EC3 Part 1-10[3], 2.2(5), (radiation loss).
(iii) $\Delta T_f = 0$. See EC3 Part 1-10[3], 2.2(5) and 2.3.1(2), (impact loading).
(iv) $\Delta T_{fcf} = 0$. See EC3 Part 1-10[3], 2.2(5) and 2.3.1(2), (cold forming).

If any of conditions (i) to (iv) are not complied with, an appropriate adjustment towards the right side of the table should be made in accordance with the relevant clauses above.

2.9 General principles

For welded details the type 'Welded – moderate' should be used except where there are welded attachments over 150mm long with a transverse weld or where there are butt welds joining the full cross section. Where the attachment does not exceed 50mm in width 'Welded – severe' should be used and 'Welded – very severe' for wider attachments. Where the butt weld is in a fabricated section 'Welded – severe' should be used and 'Welded – very severe' for butt welds in rolled sections. Where there are welds across the ends of cover plates they should be treated as attachments over 150mm long.

Table 2.7 does not include any allowance for gross stress concentrations. Where these occur reference should be made to the National Annex to EC3 Part 1-10[3]. A typical situation where stress concentrations do occur is where there are welded connections to unstiffened flanges and in welded tubular connections. As an alternative to using the National Annex to EC3 Part 1-10[3] it will usually be sufficient to treat these details as 'Welded – very severe'.

2.9.6 Prevention of lamellar tearing

Lamellar tearing is the splitting open of inclusions in a steel plate due to stresses in the through thickness direction. The stresses are generally those due to fabrication and weld shrinkage. One way of avoiding this phenomenon is to specify through thickness tensile testing of the material in accordance with BS EN 10164[27]. A Z value of Z15, Z25, or Z35 is specified depending on the through thickness ductility required.

EC3 Part 1-10[3] includes specific guidance on methods to determine whether Z grade material is required and what particular value should be used. This guidance is a Nationally Determined Parameter and the UK National Annex has replaced this with guidance in the PD to EC3 Part 1-10[26]. This guidance is set out below.

In most situations the risk of lamellar tearing is to be mitigated by fabrication control measures i.e. selection of appropriate material (e.g. low levels of sulphur), and suitable details and welding procedures. Z35 material should be specified in the following high-risk situations:
– tee joints, t_z>35mm
– cruciform joints, t_z>25mm
– corner joints without preparation of through material, t_z>20mm.

The thickness t_z for all butt welds and deep penetration welds should be taken as the thickness of the incoming material (see Figure 2.3). For fillet welds t_z should be the throat size of the largest fillet weld. Figure 2.3 also shows how, by preparing the through material at corner joints through thickness stresses are limited and the requirement for Z grade material is avoided.

General principles

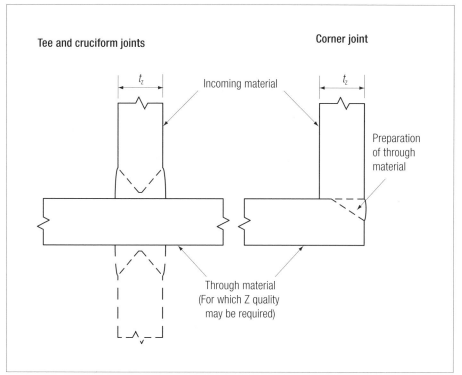

Fig 2.3 Joints where lamellar tearing is considered

2.10 Section classification

Sections are classed in the Eurocode as:
- Class 1 cross sections are those that can form a plastic hinge with the rotation capacity required for plastic analysis.
- Class 2 cross sections are those that can develop their plastic moment resistance, but have limited rotation capacity.
- Class 3 cross sections are those in which the calculated stress in the extreme compression fibre of the steel member can reach its yield strength, but local buckling is liable to prevent development of the plastic moment of resistance.
- Class 4 cross sections are those in which it is necessary to make explicit allowances for the effects of local buckling when determining their moment resistance or compression resistance. This is done using effective widths or reduced strength and methods are given in BS EN 1993-1-5:2006[28] (EC3 Part 1-5).

The limits for the various conditions are given in EC3 Part 1-1[1] in Table 5.2. Some of these are shown in Table 2.8.

Table 2.8 Selection of width to thickness ratios for classification of elements

Element	Ratio	Class 1	Class 2	Class 3
Outstand element with uniform compression	c/t	9ε	10ε	14ε
Internal element with uniform compression	c/t	33ε	38ε	42ε
Internal element with neutral axis at mid-depth	c/t	72ε	83ε	124ε
Angles under compression	h/t	Not applicable	Not applicable	15ε
	$(b+h)/(2t)$			11.5ε

Notes
a This is based on Table 5.1 of EC3 Part 1-1[1] and that table includes ratios for other sections and different stress distributions.
b The dimension c is the flat distance of the element, clear of fillets or welds. For outstand flanges it is as shown in Figure 1.2. For the web of an I section under compression or bending the dimension c is the dimension d shown on Figure 1.2.
c Both limits for the angle have to be satisfied for the section to be considered class 3.
d The symbol $\varepsilon = \sqrt{235/f_y}$

2.11 Methods of analysis

2.11.1 General approach

EC3 permits two approaches to the analysis of a structure, these are:
- Elastic global analysis, where the member moments and forces are derived from the actions by means of an elastic analysis of the frame for both ultimate and serviceability limit states.
- Plastic analysis for the ultimate limit state, where the plastic behaviour is taken into account. This may be further divided into two classes:
    - Rigid plastic methods, usually considered as simple plastic design in the UK.
    - Elastic plastic methods, where a more detailed study of the plasticity of the frame is considered.

With the exception of the analysis of portal frames, plastic analysis is not covered in this *Manual*.

EC3 Part 1-1[1] requires that the effect of the deformed geometry (second-order effects) needs to be considered in the analysis. A second order analysis can be used but in certain situations a first order analysis is acceptable. Where the effects of the deformations are small they can be neglected, where they are larger they can be included by amplifying the effects from a first-order analysis, and where they are likely to be very large a second order analysis is required. The process for establishing the sensitivity of regular buildings and the amplification factors to be used is described in Section 2.7.

2.11.2 Joint modelling

EC3 gives three types of joint model:
- Simple – in which the joint may be assumed not to transmit bending moments i.e they are nominally pinned. The joints must have sufficient rotation resistance. With this type of joint the frame will always require either bracing or shear walls of some form to resist horizontal loading.
- Semi-continuous – in which the behaviour of the joint needs to be taken into account in the analysis.
- Continuous – in which the behaviour of the joint may be assumed to have no effect on the analysis i.e. the joints are assumed to be rigid (infinitely stiff).

These models are similar to those in common use in the UK in the past. Classification of joints is given in EC3 Part 1-8[2]. Numerical methods are given but it is also permitted to classify them on the basis of previous satisfactory performance. The UK National Annex states that connections designed according to the principles given in the publication *Joints in Steel Construction-Simple Connections*[29] may be classified as nominally pinned connections. It also states that guidance in the publication *Joints in Steel Construction-Moment Connections*[30] may be used to classify connections as being rigid.

Structures with semi-continuous joints are outside the scope of this *Manual*. Guidance on the stiffness and strength of semi-continuous joints between I sections is given in EC3 Part 1-8[2]. Some guidance on the use of partial strength connections in plastically designed semi-continuous braced frameworks is given in the publication *Design of semi–continuous braced frames*.[31] Guidance on un-braced semi-continuous frames (wind moment frames) is given in the publication *Wind-moment design of low rise frames*.[32]

For frames with simple joints the structure may be taken as statically determinate. This means that beams are taken as being pinned at the supports and columns and ties are designed as simple members. However it will still be necessary to consider some moments in columns due to the eccentricity of load at connections.

Trusses may be taken as pinned or rigid, although no specific criteria are included, except in Section 5.1.5 of EC3 Part 1-8[2] where the joints are between tubular elements.

3 Braced multi-storey buildings – General

3.1 Introduction

This Section offers advice on the general principles to be applied when preparing a scheme for a braced multi-storey structure. It is assumed that simple construction will be used i.e. the joints are assumed not to transmit bending moments.

3.2 Actions (loads)

Eurocode 1 Part 1-1[12] allows reductions in imposed load based on the area or the number of floors supported. In preliminary design, to save time, these load reductions need not be taken into consideration. Where the imposed load is an accompanying action i.e. when it is multiplied by a combination factor as described in Section 2.8.1, the reduction based on the number of floors cannot be used. The load factors, γ_Q and γ_G, for use in design should be obtained from Table 2.2.

Temperature effects should also be considered where appropriate, especially when the distance between movement joints is greater than that recommended in Section 2.5.

Where forces in elements are caused by more than one variable action the load factors and directions (if relevant) should be chosen to give the maximum effect in the element. This may involve multiple loadcases to cover all the elements in a building.

Care should be taken not to underestimate the dead loads, and the following figures should be used to provide adequate loads in the absence of firm details:

floor finish (75mm screed)	1.8kN/m² on plan
ceiling and service load	0.5kN/m² on plan
movable partitions (self weight ⩽2.0kN/m)	0.8kN/m² on plan
movable partitions (self weight ⩽3.0kN/m)	1.2kN/m² on plan
blockwork partitions	2.5kN/m² on plan
external walling – curtain walling and glazing	0.5kN/m² on elevation
cavity walls (lightweight block/brick)	3.5-5.5kN/m² on elevation

Unit weight of normal weight aggregate reinforced concrete should be taken as 25kN/m^3.

Unit weight of lightweight coarse aggregate reinforced concrete should be taken as 19kN/m^3.

3.3 Material selection

For multi-storey construction in the UK, S355 steel is usually the most economic solution for beams acting compositely with the floors or where deflection does not govern the design; otherwise S275 steel may be used for beams. S275 material is typically used for plates in connections. Similar sections of differing grades of steel should not be employed in the same project, unless there is exceptional quality control to prevent the members being exchanged. For columns, S355 steel should be considered where it is intended to reduce the sizes to a minimum. Class 8.8 bolts should normally be used throughout.

3.4 Structural form and framing

3.4.1 Stability

Stability should be provided by arranging suitable braced bays or cores deployed symmetrically wherever possible to provide stability against lateral forces in two directions approximately at right-angles to each other. Suitable types of bracing are described in Section 2.3.3. Typical locations are shown in Figures 3.1 and 3.2; they will usually correspond to service cores. Where movement joints are considered necessary (see Section 2.5) the stability of each part of the structure will need to be considered separately.

The combination of the wind and equivalent horizontal forces on the structure to give the critical load cases should be assessed and divided into the number of bracing bays resisting the horizontal forces in each direction. As noted in Sections 2.3.1 and 2.6.1, there may be torsional effects on the building which will mean that the loads are not equally distributed between the bracing bays.

If the permanent stability elements cannot be constructed to suit the erection sequence of the frame, temporary bracing may be required and this should be described in the erection methods and preferably shown on the drawings.

Braced multi-storey buildings – General 3.4

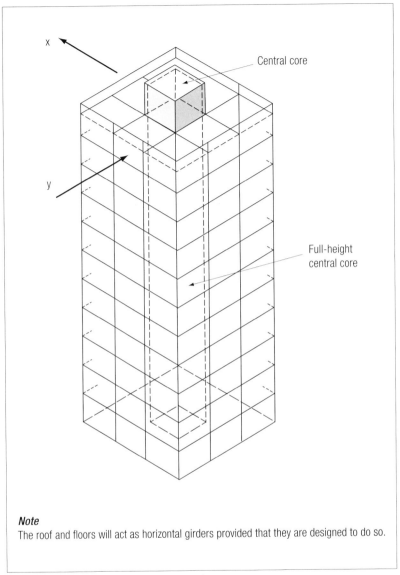

Note
The roof and floors will act as horizontal girders provided that they are designed to do so.

Fig 3.1 Braced frame square on plan – central core

3.3 Braced multi-storey buildings – General

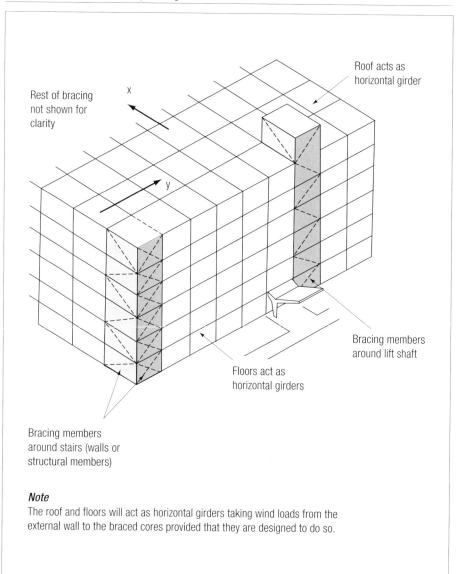

Fig 3.2   Braced frame rectangular or square on plan

3.4.2 Framing

A column grid and beam layout should be established. The framing will need to suit the architecture and services as well as structural efficiency. The beams should tie the columns together in two directions approximately at right angles. Section 3.7 gives loading requirements for horizontal and vertical ties to satisfy robustness requirements.

If there are movement joints it is preferable to provide columns each side of the joint rather than try and put a joint at the end of a beam. Where this is not possible any resultant eccentricity must be taken into account and care taken to ensure any joint remains serviceable under loading.

Typical structural forms for various spans are discussed in SCI publication *Comparative structure cost of modern commercial buildings*[33] and some of these are shown in Figure 3.3. The options shown in Figure 3.3 are generally for secondary beams, primary beams will usually be limited to 10-12m span. The Slimflor and Slimdek options are where the beams are incorporated in the depth of the slab, which is usually increased to give a uniform soffit. Deep composite decking is used with Slimdek to span from one column line to the next and avoid the need for secondary beams. Requirements for stiffer edge beams should be considered and additional façade columns provided if necessary (see Section 2.8.2.1).

The arrangement should take account of possible problems with foundations, e.g. columns immediately adjacent to site boundaries may require balanced or other special foundations.

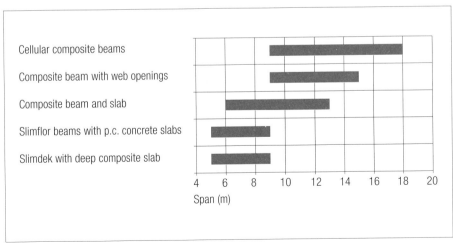

Fig 3.3 Summary of typical spans of structural forms for secondary beams

3.4 Braced multi-storey buildings – General

3.4.3 Flooring

Except for domestic construction, the floor is usually either concrete on profiled steel decking or precast concrete. Either system should be designed so that temporary propping is not required during construction. The flooring can provide stability to the top flange of beams and can also provide a diaphragm that distributes horizontal forces and stability effects. It can therefore have a significant effect on the structure.

Table 3.1 summarises the salient features of the various types of flooring commonly used in the UK.

It should be noted that the normal practice for profiled steel decking with concrete topping is to design the slab as simply supported and hence there is little control of cracking. This may not be suitable for some finishes.

Braced multi-storey buildings – General 3.4

Table 3.1 Details of typical flooring systems and their relative merits

Floor type	Typical span range (m)	Typical depth (mm)	Construction time	Degree of lateral restraint to beams	Degree of diaphragm action	Main areas of usage and remarks
Timber	2.5 – 4.0	150 – 300	Medium	Poor	Poor	Domestic
In situ concrete	3.0 – 6.0	150 – 250	Medium	Very good	Very good	All categories but not often used for multi-storey steel construction, as formwork and propping are required
Precast concrete	3.0 – 6.0	110 – 200	Fast	Bad – fair	Fair – good (detailed design can improve the performance)	All categories especially multi-storey commercial. Ensure that the joints between units are designed and executed to prevent differential movement
Profiled steel decking composite with concrete topping	Shallow decks 2.5 – 3.6m unpropped	110 – 150	Fast	Very good	Very good	All categories. A check on the effects of vibration will be required
	Deep decks 4.0 – 6.5m unpropped	200+				
	Deep decks 5.0 – 9.0m propped					

Notes
a Timber floors should be designed to EC5[34].
b *In situ* concrete floors should be designed to EC2[35].
c Precast concrete floors should be designed to EC2 and to the guides provided by the manufacturers of proprietary flooring systems. The condition during erection where there is unbalanced loading on the beam and no lateral restraint should be considered.
d Profiled-steel-decking/composite floors should be designed to EC3 and EC4 and to the literature provided by the manufacturers of the proprietary metal-decking systems.

3.5 Fire resistance

Structural steel members may require protection by insulating materials to enable them to sustain their actions during a fire. The type and thickness of insulation to be applied depends on the period of fire resistance required and the section factor of the member (heated perimeter/cross sectional area).

Fire resistance is given in Building Regulations[36] in terms of a standard fire resistance period, usually 30 mins, 1, 2 or 4 hours, depending on the nature of the building and the consequences of the failure of the structure.

EC3 Part 1-2[25] may be used to determine the fire resistance of steel structures. Further guidance is given in:
– *Structural fire safety: A handbook for architects and engineers*[37]
– *Guidelines for the construction of fire-resisting structural elements*[38].

Details of various proprietary systems may be found in the Association for Specialist Fire Protection Contractors and Manufacturers' manual on fire protection[39].

In the absence of specific information, choose a fire resistance period of 1h for the superstructure and 2h for ground floor construction over a basement and the basement structure. This may be achieved by choosing one of the alternatives in Table 3.2.

Table 3.2   Fire protection: type of protection

Type of protection	Approximate thickness in mm for period of fire resistance	
	60min	120min
Spray	20	35
Boarding	15	30
Intumescent coating (normally up to 1h)	1 – 5	–
Reinforced concrete casing – loadbearing	50	50
Reinforced concrete casing (Min. Grade 20) – non-loadbearing	25	25

More advanced methods of dealing with fire are given in two Institution publications:
– *Introduction to the fire safety engineering of structures*.[40]
– *Guide to the advanced fire safety engineering of structures*.[41]

3.6 Corrosion protection

In order to corrode, steel must be in the presence of both air and water. In a dry controlled environment no protection is necessary and this is defined in EC3 as being where the internal relative humidity does not exceed 80%. This will apply to steel for the majority of office buildings but ancillary areas (e.g. kitchens, shower rooms) may require protection. Particular uses e.g. salt stores, will require specific measures that are outside the scope of this *Manual*. EC0[16] gives the points that should be considered, but no guidance as to how the level of protection should be determined. The points given include:
- the intended or foreseeable use of the structure
- the required design criteria
- the expected environmental conditions
- the composition of the steel
- the shape of the members and structural detailing
- the particular protective measures
- the intended maintenance during the design working life.

Guidance on corrosion protection is given in BS EN ISO 12944[42] but particular guidance for building steelwork is available in CIRIA Guide 174 *New paint systems for the protection of constructional steelwork*[43]. These documents consider the exposure risk and give suggested levels of protection for periods to first maintenance. For most constructional steelwork the appropriate corrosion protection system can be found using CIRIA 174. This Guide describes seven possible specifications, two external and five internal (see Table 3.3).

The life to first maintenance relative to the environment for these specifications is given in Tables 3.4 and 3.5. This is the life to first maintenance of the coating and is intended to be such that there is some degradation of the coating but a complete reapplication of the original specification (including preparation back to bare metal) is not required. As such it is a matter of engineering judgement.

Bolts and other fittings should be protected to the same level as the main structural members. It may often be necessary to apply the protection after construction so that there is no damage to the protective layer. Any making good of the protective layer, because of damage during handling, should be to the manufacturer's specification.

Galvanizing is sometimes considered in place of the specifications shown in Table 3.3 e.g. I-3 or, where no decorative coating is required, E-1, E-2, or I-5. This can provide a damage tolerant coating but will generally require inspection of the steelwork after galvanizing. Guidance on what might be required is in the BCSA/GA publication *Galvanizing Structural Steelwork*[44].

3.6 Braced multi-storey buildings – General

Table 3.3 Specifications from CIRIA 174[43]

Location	Primer		Intermediate		Finish	
CIRIA Spec	Type	DFT[a] (μm)	Type	DFT[a] (μm)	Type	DFT[a] (μm)
External steel						
E-1	Epoxy zinc rich	75	Epoxy MIO	100-125	Acrylic/urethane	50
E-2	Epoxy zinc phosphate	75	Epoxy MIO	100-125	Acrylic/urethane	50
Internal steel						
I-1	None	–	None	–	None	–
I-2	Epoxy zinc phosphate	50	None	–	Acrylic/urethane	50
I-3	Epoxy zinc rich	50	None	–	None	–
I-4	Epoxy primer/finish	125	None	–	None	–
I-5	Epoxy zinc phosphate	75	Epoxy MIO	100-125	Acrylic/urethane	50

Note
a DFT is the dry film thickness.

Table 3.4 Life to first maintenance for external specifications

Specification	Environment		
	Rural	Urban/industrial	Coastal
E-1	25 years	15-20 years	15-20 years
E-2	20 years	15-20 years	15-20 years

Table 3.5 Environments and life to first maintenance for internal specifications

Specification	Environment	Life to first maintenance
I-1	Internal steel in controlled or air-conditioned spaces	Life of building
I-2	Internal environment with decorative requirement	Nominally 5-10 years
I-3	Steel in cavity walls in clear separation outside the vapour barrier of the cladding system	Life of building
I-4	Semi-controlled environment: Plant rooms, dry warehouses, roof voids (where occasional condensation may occur)	15-20 years
I-5	Uncontrolled environments that are frequently damp or wet: swimming pools, kitchens, laundries etc.	15-20 years

Note
The descriptions of environment and values of life to first maintenance are those given in CIRIA 174[43].

3.7 Robustness

3.7.1 Introduction

Multi-storey construction that has been framed in accordance with the recommendations given in Section 3.4 and designed in accordance with the rest of the *Manual*, should produce a robust construction but BS EN 1991-1-7: 2006[45] has more particular guidance. Where the consequence of failure is significant, it states that one of the following approaches should be adopted to mitigate the potential failure from accidental situations:
- Designing key elements, on which the stability of the structure depends, to sustain a particular accidental load. The UK National Annex has accepted the recommended value of $34kN/m^2$.
- Designing the structure so that in the event of localised failure the stability of the structure (or a significant part of it) would not be endangered. The UK National Annex has accepted the recommended limits for localised failure. These are a minimum of $100m^2$ or 15% of the floor area on two adjacent floors.
- Applying prescriptive design/detailing rules e.g. tying.

More detail is given in informative Annex A[45], which has been accepted in the UK National Annex. This gives information on the classification of buildings and gives particular information on the third approach described above. Table 3.6 shows the classes and the basic requirements. The examples in Table 3.6 are very limited, additional examples are given in the Annex. The alternative approach for class 2b buildings and the risk assessment for class 3 buildings are outside the scope of this *Manual*.

Table 3.6 Building classification for robustness

Class	Example	Requirements
1	Agricultural buildings Single occupancy houses not exceeding 4 storeys	None
2a Lower risk group	Offices not exceeding 4 storeys	Effective horizontal ties
2b Upper risk group	Offices exceeding 4 storeys	Effective horizontal and vertical ties, or alternative approach by considering notional removal of elements
3	Offices exceeding 15 storeys	A risk assessment is required

3.7 Braced multi-storey buildings – General

3.7.2 Horizontal ties

The requirement for horizontal ties is given in Annex A to EC1 Part 1-7[45]. They should be provided around the perimeter of each floor and roof level and internally in two right angle directions to tie the columns securely to the rest of the structure of the building. The ties should be continuous and be arranged as closely as practicable to the edges of floors and lines of columns. At least 30% of the ties should be located within the close vicinity of the grid lines of the columns. It should be noted that columns will also be required to be restrained as described in Section 2.6.2.2 and this requirement may provide all or part of the tying force.

The design load for the ties depends on the vertical load in the accidental situation and is equal to the following values:

For internal ties, $T_i = 0.8(g_k + \psi_1 q_k)sL$ or 75kN, whichever is the greater (A.1 EC1-1-7)

For perimeter ties, $T_i = 0.4(g_k + \psi_1 q_k)sL$ or 75kN, whichever is the greater (A.2 EC1-1-7)

where:
s is the spacing of the ties
L is the span
g_k is the permanent load on the floor or roof
q_k is the variable load on the floor or roof
ψ_1 is the frequent combination factor from EN 1990[16].

The factor ψ_1 is that for frequent combinations and usually has a value of 0.5. Exceptions are:
– congregation, traffic and shopping areas, $\psi_1 = 0.7$
– storage areas, $\psi_1 = 0.9$
– snow (buildings less than 1000m above sea level), $\psi_1 = 0.2$
– wind loads, $\psi_1 = 0.2$

The tie force can be taken in steel sections, reinforcement or decking which is anchored by shear studs to the steel structure. A combination of these may be used. Such a load should not be assumed to act simultaneously with other loads on the structure.

3.7.3 Vertical ties

Annex A to EC1 Part 1-7[45] also gives the requirements for vertical ties. Each column should be tied continuously from the foundations to the roof. They should be capable of resisting a tie force equal to the largest design reaction applied to the column from any one storey. Such a load should not be assumed to act simultaneously with other loads on the structure.

3.8 Connections

It is important to establish the general form and type of connections assumed in the design of the members and to check that they are practicable. It is also important to consider the location of edge beams and splices, the method and sequence of erection of steelwork, access and identification of any special problems and their effects on connections such as splicing/connection of steelwork erected against existing walls. The extent of welding should also be decided, as it may have an effect on cost. Preliminary design of typical connections is necessary when:
– appearance of exposed steelwork is critical
– primary and secondary stresses occur that may have a direct influence on the sizing of the members, e.g. local wall strengthening in tubular connections
– connections are likely to affect finishes such as splices affecting column casing sizes and ceiling voids
– steelwork is connected to reinforced concrete or masonry, when greater constructional tolerances may be required, which can affect the size and appearance of the connections
– unusual geometry or arrangement of members occurs
– holding bolts and foundation details are required
– a detail is highly repetitive and can thus critically affect the cost.

The normal execution standards and specifications require a degree of checking on bolts and welding. Where there are critical connections, the need for more particular inspection e.g. NDT of welds should be considered.

Guidance on the design of connections is given in Sections 10 and 11.

4 Beams

4.1 Non-composite beams with full lateral restraint

4.1.1 Introduction

Section 4.1 applies to beams where the frictional force or positive connection between the compression flange of the member and the floor it supports is sufficient to restrain the beam against buckling. Guidance on the force required to be resisted is given in Section 2.6.2.3. In practice a system capable of resisting a lateral force of at least 2.5% of the maximum force in the compression flange arising from the design loads should usually comply. Appendix A gives values of moment resistance for UB sections.

4.1.2 Design procedure

(1) Calculate the design load. Assuming that the only variable action that needs to be considered is the imposed floor load then this is equal to 1.5 × imposed load + 1.35 × dead load. Then calculate the maximum design bending moment M_{Ed}, and the design shear forces V_{Ed}.

(2) Choose a section such that its moment of resistance $M_{y,Rd}$ about its major axis $\geqslant M_{y,Ed}$

In order to choose a trial section that will not be critical in local buckling, it is necessary to note that elements and cross-sections have been classified according to their behaviour with regard to local buckling. Descriptions of the section classes are given in Section 2.10.

The limiting width/thickness ratios of the elements of the sections stated in Table 5.2 of EC3 Part 1-1[1] should be consulted for the determination of the class of section. In order to assist the selection of suitable sections for use as beams in bending the classifications are given in Table 4.1.

Beams   4.1

Table 4.1   Section classification for bending only

The following sections are class 3 (semi-compact), all other UB and UC sections are either class 1 (plastic) or class 2 (compact):	
Grade S275 steel	Grade S355 steel
152 × 152 × 23 UC	152 × 152 × 23 UC 305 × 305 × 97 UC 356 × 368 × 129 UC
Notes a The Corus Advance range of sections includes UB and UCs that are not in BS4[46], these are included in the above. b Dimensions of all sections in the Advance range are given in SCI publication P363[6].	

The value of the moment resistance $M_{y,Rd}$ of a beam about its major axis may be determined from:

$M_{y,Rd} = W_{pl,y} f_y / \gamma_{M0}$ for class 1 and 2 cross sections, and

$M_{y,Rd} = W_{el,min,y} f_y / \gamma_{M0}$ for class 3 cross sections

where:
$W_{pl,y}$ is the plastic modulus of the section about the major axis
$W_{el,min,y}$ is the minimum elastic modulus of the section about the major axis
f_y is the design strength of the steel obtained from Table 2.6 according to the steel grade and flange thickness
γ_{M0} is the partial material factor with a value of 1.0 according to the UK National Annex.

Where the shear on the section is more than 50% of the shear resistance (see below), then the moment resistance must be reduced. It will be found that this will only be critical if there are heavy point loads near a support, or continuous members or cantilevers. Details of the calculations will be found in EC3 Part 1-1[1] clause 6.2.8.

If there are bolt holes in the flanges at a point of high moment then the moment resistance of the section may be reduced, see EC3 Part 1-1[1] clause 6.2.2.2. Provided the net area of the flange is greater than 82% of the gross area there should be no reduction. This is based on the partial safety factor γ_{M2} being 1.1 as described in Section 2.9.1.

(3) Check the deflection satisfies the several deflection limitations described in Section 2.8.3.1 from the second moment of area *I*.
For the simply supported beams covered in this *Manual*:

4.1 Beams

The value of $\hat{\delta}$ for a uniform load may be taken as:

$$\hat{\delta} = \frac{6.2FL^3}{I}\,\text{mm}$$

When the beam carries a central point load, the following equation may be employed:

$$\hat{\delta} = \frac{9.9FL^3}{I}\,\text{mm}$$

When more than one load is imposed on the beam the principle of superposition may be used for each point load in turn. The determination and use of the central deflection is normally sufficiently accurate for practical design.

where:
$\hat{\delta}$ is the maximum deflection for limit being considered (mm)
F is the characteristic (Unfactored) point or total load (action) (kN)
L is the span (m)
I is the second moment of area of the section (cm^4).

For cantilevers with backspans and continuous beams the deflections should be calculated from first principles taking into account the slopes at the supports and the ratio of the length of the cantilever to the span of its adjoining member.

(4) Check the shear resistance of the selected beam.

If the ratio of the web depth to the web thickness is greater than 72ε, where ε is defined in Table 2.8, the shear buckling resistance of the section must be checked in accordance with BS EN 1993-1-5:2006[28] (EC3 Part 1-5). This is not necessary for any UB or UC section.

The plastic shear resistance $V_{pl,Rd}$ is given by the following equation:

$$V_{pl,Rd} = A_v\left(f_y/\sqrt{3}\right)/\gamma_{M0} \qquad \text{(6.18 EC3-1-1)}$$

where:
A_v is the shear area of the section
f_y is the design strength of the material

EC3 Part 1-1[1] Clause 6.2.6 gives expressions for A_v for various sections. For I and H sections A_v can usually be taken as ht_w. This is conservative for all UB and UC sections except for the three heaviest UC sections (467, 551 and 634kg/m) where it overestimates the shear area by up to 5%.

4.2 Non-composite I and H section beams without full lateral restraint

4.2.1 Introduction

These sections may fail by means of lateral torsional buckling.

Section 4.2 applies to elements with the bending taking place about the strong (y-y) axis:
– as beams with a load at or between points of restraint
– as columns subject to bending moments.

It applies to I and H sections with equal flanges though guidance for other sections is given in Section 4.3 It has been assumed that the forces and actions are applied through the shear centre of the section and that the sections are of classes 1, 2 or 3. For all other cases see EC3.

All unrestrained beams must satisfy all the requirements for restrained beams as well as complying with these additional rules.

Appendix A gives values of buckling resistance for various effective lengths for UB sections.

4.2.2 Design procedure

(1) Calculate the maximum design moments and shears on the section and the second moment of area as for a restrained section.

(2) Select a trial section.

(3) For the selected section determine the value of $\bar{\lambda}_{LT}$ the relative slenderness (see Section 4.2.3)

(4) Look up the value of χ_{LT} from Figure 4.1 using the appropriate curve for the section h/b. Where there is non-uniform moment, EC3 Part 1-1[1] gives a method whereby the value can be modified to give a lower slenderness.

(5) Calculate the resistance moment from:

$$M_{b,Rd} = \chi_{LT}(W_y f_y)/\gamma_{M1} \qquad (6.55 \text{ EC3-1-1})$$

where:
W_y is $W_{pl,y}$ for class 1 and 2 cross sections and $W_{el,min,y}$ for class 3 cross sections.

4.2 Beams

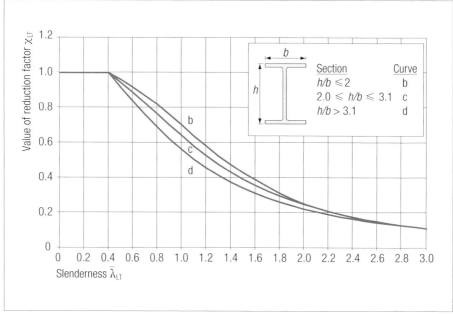

Fig 4.1   Graph of reduction factor χ_{LT} from slenderness $\bar{\lambda}_{LT}$ for rolled sections

If this is greater than or equal to the maximum moment on the part of the member being considered then, providing the restrained beam checks are satisfied and there are no other unrestrained lengths the section may be taken as adequate. If there are other unrestrained lengths then these must be checked in the same way as before.

4.2.3 Calculation of relative slenderness

EC3 Part 1-1[1] gives the following expression for the relative slenderness:

$$\bar{\lambda}_{LT} = \sqrt{\frac{W_y f_y}{M_{cr}}}$$

In this expression, M_{cr} is the elastic critical buckling moment and the code does not give any guidance on calculating this value. Direct methods of calculating the relative slenderness are given below.

The relative slenderness $\bar{\lambda}_{LT}$ may conservatively be calculated from the expressions in Table 4.2.

Beams 4.2

Table 4.2   Simplified expressions for relative slenderness

Grade S275	Grade S355
$\bar{\lambda}_{LT} = \dfrac{[L/i_z]}{96}$	$\bar{\lambda}_{LT} = \dfrac{[L/i_z]}{85}$

A less conservative value can be calculated using the following expression:

$$\bar{\lambda}_{LT} = \frac{1}{\sqrt{C_1}} UVD \bar{\lambda}_z \sqrt{\beta_w}$$

where:
C_1 is a factor based on the moment diagram between the points of lateral restraint, a value of $C_1=1.0$ is conservative. For cantilevers C_1 should be taken as 1.0. For other beams Table 4.3 shows the effect on slenderness of certain moment diagrams. The factors in Table 4.3 assume that the load is not destabilising.
U is a parameter dependent on the section geometry and can conservatively be taken as 0.9 or as:

$$U = \sqrt{\frac{W_{pl,y}}{A} \sqrt{\frac{I_z}{I_w}}}$$

V is a parameter related to the slenderness and may be taken conservatively as 1.0 or from Table 4.4.
β_w is the ratio $W_y/W_{pl,y}$ where W_y is defined in Section 4.2.2.
D is a parameter to allow for the effect of destabilising loads i.e. when the load is applied at the top flange and both the load and top flange are free to move sideways. Values are given in Tables 4.5 and 4.6. Where the load is destabilising C_1 should be taken as 1.0.

$$\bar{\lambda}_z = \frac{\lambda_z}{\lambda_1}$$

where: $\lambda_z = \dfrac{kL}{i_z}$, and $\lambda_1 = \pi \sqrt{\dfrac{E}{f_y}} = 93.9\varepsilon$ and $\varepsilon = \sqrt{\dfrac{235}{f_y}}$

λ_1 can be taken as 86.8 for S275 steel and 76.4 for S355 steel.

The parameter k depends on the restraint conditions at supports and values are given in Tables 4.5 and 4.6. It should be noted that Table 4.3 is only valid for $k=1.0$.

L is the distance between restraints to the compression flange or for a cantilever it is the length of the cantilever and i_z is the minor axis radius of gyration of the section.

Background to the above expressions is given in the Access Steel documents SN002a[47] and SN009a[48].

4.2 Beams

Table 4.3 Determination of C_1

Loading and support conditions	Bending moment diagram	$1/(C_1)^{0.5}$
Simply supported beam with UDL		0.94
Fixed-ended beam with UDL		0.62
Simply supported beam with central point load		0.86
Fixed-ended beam with central point load		0.77
Beam with end moments M and ψM	Or	See Figure 4.2

Note
For point loads at third points the values for uniform loads can be used.

Beams 4.2

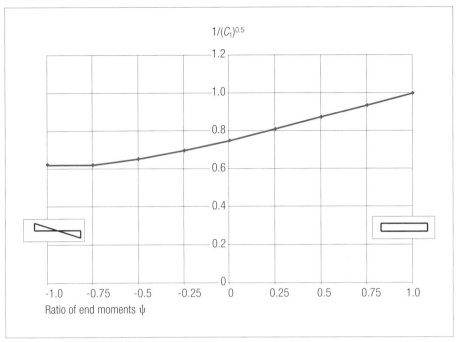

Fig 4.2 Variation of $1/(C_1)^{0.5}$ with ψ

Table 4.4 Values of slenderness parameter V

λ_z									h/t_f
	5	10	15	20	25	30	35	40	45	50
30	0.77	0.91	0.96	0.97	0.98	0.99	0.99	0.99	0.99	1.00
50	0.64	0.82	0.90	0.93	0.96	0.97	0.98	0.98	0.99	0.99
75	0.53	0.72	0.82	0.88	0.91	0.93	0.95	0.96	0.97	0.97
100	0.47	0.64	0.75	0.82	0.86	0.90	0.92	0.93	0.95	0.96
125	0.42	0.58	0.69	0.76	0.82	0.86	0.88	0.91	0.92	0.93
150	0.38	0.53	0.64	0.72	0.77	0.82	0.85	0.88	0.90	0.91
175	0.36	0.50	0.60	0.67	0.73	0.78	0.82	0.85	0.87	0.89
200	0.33	0.47	0.56	0.64	0.70	0.75	0.79	0.82	0.84	0.86
225	0.31	0.44	0.53	0.61	0.67	0.72	0.76	0.79	0.82	0.84
250	0.30	0.42	0.51	0.58	0.64	0.69	0.73	0.76	0.79	0.82
275	0.28	0.40	0.49	0.56	0.61	0.66	0.70	0.74	0.77	0.79
300	0.27	0.38	0.47	0.53	0.59	0.64	0.68	0.72	0.75	0.77

4.2 Beams

Table 4.5   Parameters k and D for simple beams without intermediate restraint

Conditions of restraint at supports		Parameters	
		k	D
Compression flange laterally restrained: Nominal torsional restraint against rotation about longitudinal axis	Both flanges fully restrained against rotation on plan	0.70	1.2
	Compression flange fully restrained against rotation on plan	0.75	1.2
	Both flanges partially restrained against rotation on plan	0.80	1.2
	Compression flange partially restrained against rotation on plan	0.85	1.2
	Both flanges free to rotate on plan	1.00	1.2

4.2.4 Beams with intermediate restraint

Where a beam has effective intermediate restraints the moment resistance can be based on the length between restraints using an effective length parameter k of 1.0. Where the load is destabilising the factor D should be taken as 1.2. Where load, other than self weight, is applied between the restraints the factor C_1 should be taken as 1.0. This also applies to cantilevers with intermediate restraints provided the support has moment continuity with lateral and torsional restraint. Where the load is applied at the points of restraints, C_1 can be calculated using the values in the last row of Table 4.3 for the length between restraints.

Beams   4.2

Table 4.6 Parameters k and D for a cantilever without immediate restraint

At support	At tip	k	D
Continuous with lateral restraint only	Free	3.0	2.5
	Laterally restrained on top flange only	2.7	2.8
	Torsionally restrained only	2.4	1.9
	Laterally and torsionally restrained	2.1	1.7
Continuous with lateral and torsional restraint	Free	1.0	2.5
	Laterally restrained on top flange only	0.9	2.8
	Torsionally restrained only	0.8	1.9
	Laterally and torsionally restrained	0.7	1.7
Built-in laterally and torsionally	Free	0.8	1.75
	Laterally restrained on top flange only	0.7	2.0
	Torsionally restrained only	0.6	1.0
	Laterally and torsionally restrained	0.5	1.0

Braced laterally in at least one bay — Top flange restraint

Face beams extending over several bays — Torsional restraint

Braced laterally in at least one bay — Lateral and torsional restraint

Note
If a bending moment is applied at its tip, k should be increased by 30% or by 0.3, whichever is the greater.

4.3 Other sections without full lateral restraint

4.3.1 General

The method described above can also be used for other sections with certain amendments. Certain parameters have to be changed and for sections that are not rolled, the UK National Annex requires that different curves for the slenderness reduction factor must be used. These curves are shown in Figure 4.3. They are identical to the curves for flexural (column) buckling.

4.3.2 Rolled channel sections subject to bending about the major axis

Where rolled channels are subject to bending about the major axis and they are not restrained, care needs to be taken to ensure that loads and reactions are applied to the shear centre. This is usually accommodated by making sure there is a torsional restraint where the load is applied as well as at the supports.

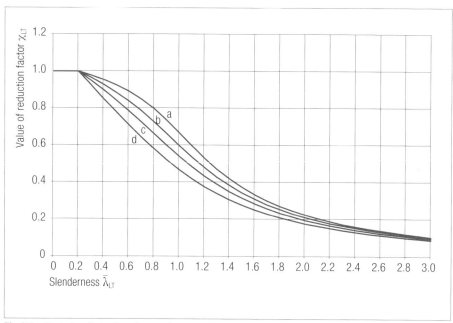

Fig 4.3   Reduction factor for other sections

Beams 4.3

Rolled channel sections that are not class 4 can be designed using the methods in Section 4.2 with the following amendments:
- If U is not calculated using the expression in Section 4.2.3 (or from published tables[6]) it shall be taken as 1.0.
- V shall be taken as 1.0 or evaluated using the expression below:

$$V = \frac{1}{\sqrt[4]{1 + \frac{1}{20}\left(\frac{\lambda_z}{x}\right)^2}}$$

The factor x may be found in published tables or can be calculated from the expression:

$$x = 1.132 \sqrt{\left(\frac{AI_w}{I_z I_t}\right)}$$

- Curve d in Figure 4.3 must be used.

4.3.3 Welded I or H sections with equal flanges

Welded doubly symmetric I or H sections that are not class 4 can be designed using the methods in Section 4.2 but with the following amendments:
- Values for U and V shall be as given in Section 4.2.3.
- Figure 4.3 must be used instead of Figure 4.1 with curve c for sections with $h/b \leqslant 2$ and curve d for other sections.

Note The UK National Annex to EC3 Part 1-1[1] limits the applicability of clause 6.3.2.3 of EC3 Part 1-1[1] to sections with $h/b \leqslant 3.1$. However the NA requires that the buckling curves for these sections are identical to those for the more general clause 6.3.2.2 and this clause has no limit. Despite this it is recommended that care be taken with such sections as deformation of the cross section can reduce the buckling resistance.

4.3.4 Welded doubly symmetrical box sections

Welded doubly symmetrical box sections that are not class 4 can be designed using the methods in Section 4.2 with the following amendments:

- relative slenderness $\overline{\lambda}_{LT} = \sqrt{\frac{W_y f_y}{M_{cr}}}$

- elastic critical moment $M_{cr} = C_1 \frac{\pi}{kL} \sqrt{EI_z GI_t}$

- Figure 4.3 must be used instead of Figure 4.1 with curve c for sections with $h/b \leqslant 2$ and curve d for other sections.

4.3 Beams

4.3.5 Rolled hollow sections

Rolled hollow sections complying with BS EN 10210[23] or BS EN 10219[24] are rarely susceptible to lateral torsional buckling. Rectangular hollow sections with h/b less than or equal to 2 and a relative slenderness $\bar{\lambda}_z$ less than 3.9ε ($\lambda_z < 370\varepsilon^2$) do not need to be checked for lateral torsional buckling.

If there is a need to check lateral torsional buckling the method in Section 4.2 can be used with the following amendments:

– relative slenderness $\bar{\lambda}_{LT} = \sqrt{\dfrac{W_y f_y}{M_{cr}}}$

– elastic critical moment $M_{cr} = C_1 \dfrac{\pi}{kL} \sqrt{EI_z GI_t}$

– Figure 4.1 can be used but for cold-formed sections curve c must be used in place of curve b and curve d in place of curve c.

4.3.6 Plates and flats

Plates and flats of height h and thickness t can be designed using the methods in Section 4.2 for class 1 or 2 sections with the following amendments:

– relative slenderness $\bar{\lambda}_{LT} = 0.877 \sqrt{\dfrac{f_y L h}{E t^2}}$

– curve d of Figure 4.3 must be used.

4.3.7 Tee sections

Rolled tee sections can be designed using the methods in Section 4.2 with the following amendments:

For tees cut from UC sections lateral torsional buckling need only be considered for moments about an axis in the plane of the web and the buckling resistance should be calculated by treating the section as a plate equal to the flange dimensions.

For tees cut from UB sections the moment of inertia about an axis in the plane of the flange is usually greater than that about an axis in the plane of the web. For these sections lateral torsional buckling need only be considered for moments about an axis in the plane of the flange (y-y axis). In this case the following can be used:

– relative slenderness $\bar{\lambda}_{LT} = UVD\bar{\lambda}_z \sqrt{\beta_w}$

where:
The parameter U is given in published tables[6].

Beams 4.3

The parameter V is given by the expression:

$$V = \frac{1}{\sqrt{\left(\sqrt{(w + 0.05(\lambda_z/x)^2 + \psi^2)} + \psi\right)}}$$

where:

$$w = \frac{4I_w}{I_z(h - 0.5t_f)^2}$$

The monosymmetry index ψ, the factor x and the warping constant I_w can be obtained from published tables[6] and other variables are as defined in Section 4.2. The monosymmetry index ψ should be taken as positive when the flange of the tee is in compression and negative when it is in tension. When the flange is in tension it can conservatively be taken as -1.0.

The reduction factor for the moment resistance should be obtained using curve d of Figure 4.1.

4.3.8 Rolled angles

When considering bending of unrestrained rolled angles the moments must be resolved into the directions of the principal axes u-u and v-v. The moment resistance about the v-v axis can be calculated assuming the angle is restrained. The moment resistance about the u-u axis must allow for lateral torsional buckling as follows:

The relative slenderness is given by the expression:

$$\bar{\lambda}_{LT} = 0.72 v_a \sqrt{\frac{f_y}{E}} \phi_a \lambda_v$$

The slenderness λ_v is that about the minor axis.

The equivalent slenderness coefficient ϕ_a can be found in published tables[6] and is given by the expression:

$$\phi_a = \sqrt{\frac{W_{el,u,min}^2 g}{AI_t}}$$

This should be calculated using the smallest elastic section modulus about the major principal axis $W_{el,u,min}$, the St Venant Torsion constant I_t, the area A and the factor g. The factor g allows for the beneficial effect of the deflection of non-precambered members. It can conservatively be taken as one or can be calculated using the expression:

$$g = \sqrt{\left(1 - \frac{I_v}{I_u}\right)}$$

4.3 Beams

For equal angles $v_a = 1$, for unequal angles it is given by the expression:

$$v_a = \frac{1}{\sqrt{\left(\sqrt{1 + \left(\frac{4.5\psi_a}{\lambda_v}\right)^2}\right) + \frac{4.5\psi_a}{\lambda_v}}}$$

The monosymmetry index ψ_a can be obtained from published tables[6] and λ_v is the slenderness about the minor axis.

The reduction factor for the moment resistance should be obtained using curve d of Figure 4.1.

A simple interaction equation should then be used:

$$\frac{M_{v,Ed}\gamma_{M0}}{W_v f_y} + \frac{M_{u,Ed}\gamma_{M1}}{\chi_{LT}(W_{el,u,min}f_y)} \leqslant 1$$

4.3.9 General case

Provided that they are not class 4, other sections can be designed using the methods given in Section 4.2 with the following amendments:

- relative slenderness $\bar{\lambda}_{LT} = \sqrt{\frac{W_y f_y}{M_{cr}}}$

- elastic critical buckling moment M_{cr} is calculated taking account of the position of the restraints, any non-symmetry of the cross section, the position of the applied load on the cross section and the shape of the moment diagram
- for rolled sections Figure 4.1 can be used, but for welded sections Figure 4.3 must be used. In both cases curve d shall be used.

Beams 4.4

4.4 Composite beams

4.4.1 Introduction

The advice in this *Manual* is limited to simply supported composite beams using I or H sections with equal flanges and having class 1 or 2 cross sections. The steel beam is assumed to be uniform throughout the span. The advice is based on Eurocode 4 Part 1-1. Where lightweight concrete is used it is assumed to be density class 1.8 with an oven dry density of 1800kg/m^3. The oven dry density is a value used in EC2[35] to calculate certain properties. For typical reinforced concrete of this class, EC2 suggests a density of 1950kg/m^3 for the calculation of self weight and permanent load (see Section 3.2). The slab is assumed to be supported on the top flange of the steel beam. The slab is assumed to be cast on metal deck or to be a solid slab.

4.4.2 Construction case

It is normal practice to try to avoid propping the beams during construction. The weight of the wet concrete therefore has to be taken by the steel beam and this loadcase needs to be checked. As well as the weight of the concrete a construction load needs to be considered. This is specified in BS EN 1991-1-6:2005[49]. It can be taken as a uniform load of 0.75kN/m^2 plus an additional 0.75kN/m^2 over a 3x3m square, though industry guidance may become available suggesting a lower value can be used. As required by this Eurocode, the weight of the concrete and decking as well as the construction loads are treated as variable loads and according to the UK National Annex to EC0[16] they are subject to a load factor of 1.5. Where there is metal decking that spans onto the beam this usually provides restraint. In other cases the restraint of the steel beam and the possibility of lateral torsional buckling should be considered.

4.4.3 Properties of concrete and reinforcement

The properties of concrete are taken from EC2[35]. EC4[4] uses the characteristic cylinder strength of concrete f_{ck}. The cube strength has traditionally been used to classify concrete in the UK. BS 8500[50] and BS EN 206-1:2000[51] use a strength class that has both the cube and cylinder strength. Values of f_{ck} and modulus of elasticity for various strength classes are given in Tables 4.7 and 4.8. It should be noted that the cylinder strength that corresponds to a particular cube strength is different depending on whether the concrete is normal or lightweight. Typical grades used with composite beams are C25/30, C30/37, or LC30/33.

4.4 Beams

Table 4.7 Properties of normal weight concrete

Property	Strength class						
	C20/25	C25/30	C28/35[a]	C30/37	C32/40[a]	C35/45	C40/50
f_{ck} (N/mm^2)	20	25	28	30	32	35	40
f_{cu} (N/mm^2)	25	30	35	37	40	45	50
E_{cm}[b] (kN/mm^2)	30	31	32	33	34	34	35

Notes
a Concrete class is not cited in Table 3.1, Eurocode 2 Part 1-1[35].
b Mean secant modulus of elasticity at 28 days for concrete with quartzite aggregates.

Table 4.8 Properties of lightweight concrete

Property	Strength class				
	LC20/22	LC25/28	LC30/33	LC35/38	LC40/44
f_{lck} (N/mm^2)	20	25	30	35	40
f_{lcu} (N/mm^2)	22	28	33	38	44
E_{lcm}[a] (kN/mm^2)	20	21	22	23	24

Note
a Mean secant modulus based on an air dry density of 1800kg/m^3.

Reinforcement and welded fabric (mesh) is specified to BS 4449[52] and BS 4483[53]. The characteristic yield strength of reinforcement f_{sk} can be taken as 500N/mm^2.

The partial safety factor for the concrete strength in the ultimate limit state γ_C is 1.5 and for reinforcement γ_S is 1.15. EC4 uses a design strength of concrete f_{cd} that is taken as f_{ck}/γ_C.

4.4.4 Effective concrete width

The width of concrete on either side of the steel beam that is considered to act compositely with the steel beam can be taken as the minimum of the following:
– one eighth of the span of the beam
– half the distance to any adjacent parallel beam
– when adjacent to an edge parallel to the beam, the distance to the edge of the slab.

4.4.5 Vertical shear resistance

The shear resistance of the beam is calculated using only the steel section.

4.4.6 Bending resistance

Provided the section is class 1 or 2 (see Table 4.1) the design bending resistance of the composite beam may be calculated using rigid plastic theory where the tensile strength of the concrete is neglected. Any profiled sheeting in compression is ignored. Figure 4.4 shows possible stress distributions in the concrete and steel. The bending resistance of a beam needs to be checked at critical sections, these are the point of maximum moment and sections subject to concentrated loads. Where the moment resistance of the composite beam is more than 2.5 times that of the non-composite steel beam, the composite beam should also be checked midway between any critical sections and between critical sections and the supports.

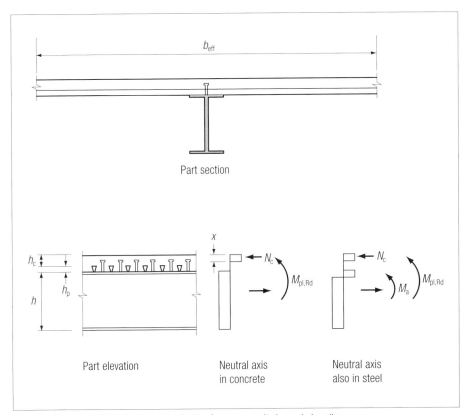

Fig 4.4 Examples of plastic stress distribution for a composite beam in bending

4.4 Beams

The design bending resistance $M_{pl,Rd}$ can be calculated as follows:

(1) N_c is the force in the concrete and is the minimum of R_c, $N_{pl,a}$ and $n_s P_{Rd}$
where:
R_c is the axial resistance of the concrete = $0.85 f_{cd} b_{eff} h_c$
$N_{pl,a}$ is the axial resistance of the steel section = $A f_y / \gamma_{M0}$
n_s is the smallest number of shear connectors between either support and the section being considered
P_{Rd} is the design resistance of the shear connectors (including for any reduction factor due to metal decking).
Minimum values of $n_s P_{Rd}$ are required as explained in Section 4.4.8.

(2) The dimension x is the depth of the concrete stress block = $N_c / (0.85 f_{cd} b_{eff})$. It can conservatively be taken as h_c.

(3) (i) If $N_c = N_{pl,a}$ then $M_{pl,Rd} = N_c (0.5h + h_p + h_c - 0.5x)$

(ii) If $N_c < N_{pl,a}$ then $M_{pl,Rd} = N_c (0.5h + h_p + h_c - 0.5x) + M_a$

where: M_a is the plastic moment resistance of the steel section reduced to account for the presence of the axial tension N_c

(iii) Where $N_c \leqslant 0.25 N_{pl,Rd}$ and $N_c \leqslant \dfrac{0.5 h_w t_w f_y}{\gamma_{M0}}$

then M_a may be taken as $M_{y,Rd}$ i.e. the full plastic moment resistance of the steel section about the y-y (major) axis

(iv) Otherwise the following approximation may be used:

$$M_a = M_{Rd}(1-n)/(1-0.5a) \text{ but } M_a \leqslant M_{y,Rd}$$

where:
$n = N_c / N_{pl,Rd}$
$a = (A - 2bt_f)/A$ but $a \leqslant 0.5$

Figure 4.5 shows the increase in moment resistance for a composite beam compared to the steel section for typical slab dimensions. The points on the graph are calculated for UB sections greater than 200mm deep from the Corus Advance range based on the dimensions and properties shown in the diagram on the Figure.

Beams 4.4

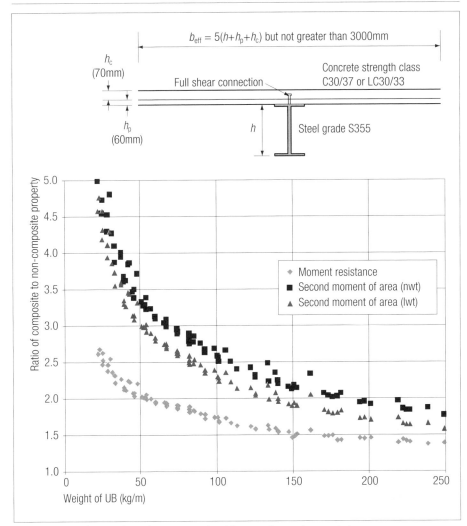

Fig 4.5 Effect of composite action on moment resistance and second moment of area

4.4.7 Shear connectors

4.4.7.1 General

The slab is normally connected to the steel beam by shear studs welded to the beam. Studs are generally type SD1 to BS EN ISO 13918[69], 19mm diameter and at least 76mm in height and these values are assumed in the resistances given in this *Manual*. The flange thickness of the beam should be at least 40% of the diameter of the stud unless it is certain that they are

4.4 Beams

placed directly above the web. To allow decking to be butt jointed on top of a beam the flange should be at least 125mm wide.

The characteristic resistance of the studs is shown in Table 4.9. According to the UK National Annex the partial safety factor to be used with these values is 1.25, unless shear stud resistances given in non-contradictory, complementary information would justify the use of an alternative value. The design value of resistance is therefore normally 80% of the characteristic value. Values based on NCCI are explained in Section 4.4.7.3.

Table 4.9 Characteristic resistance of studs

Normal weight concrete	Strength class	C20/25	C25/30	C28/35	C30/37	>C30/37
	Stud strength (kN)	81.0	92.9	99.6	102.1	102.1
Lightweight concrete	Strength class	LC20/22	LC25/28	LC30/33	LC35/38	LC40/44
	Stud strength (kN)	66.3	76.0	85.0	93.5	101.7

4.4.7.2 Resistance of headed studs with profiled steel sheeting from EC4 Part 1-1

Where the floor slab is metal decking there are reduction factors on the stud design resistance. Where the decking is parallel to the beam the reduction factor is k_l and where it is perpendicular to the beam it is k_t. Expressions for these factors are:

$$k_l = 0.6 \frac{b_0}{h_p} \left(\frac{h_{sc}}{h_p} - 1 \right) \leq 1.0 \qquad (6.22 \text{ EC4-1-1})$$

$$k_t = \frac{0.7}{\sqrt{n_r}} \frac{b_0}{h_p} \left(\frac{h_{sc}}{h_p} - 1 \right) \leq k_{t,max} \qquad (6.23 \text{ EC4-1-1})$$

The symbol n_r is the number of studs per trough, h_{sc} is the nominal height of the studs and the other symbols are shown in Figure 4.6 and Figure 4.7. The expression for k_t is not valid for more than two studs per trough or for decks where $h_p > 85$mm or for decks where the width $b_0 \leq h_p$. The expression assumes that the studs are placed centrally in the trough. If, due to stiffeners in the profile, this cannot be done, the studs should be alternately placed either side throughout the length of the beam. The upper limit $k_{t,max}$ depends on the number of studs and thickness of sheet. For sheets not exceeding 1mm thick it is 0.85 for one stud and 0.7 for two. For thicker sheets these values increase to 1.0 and 0.8 respectively.

The nominal height of the studs h_{sc} should be at least $2d$ (38mm) greater than the depth of the decking h_p. There should also be 30mm clear between the underside of the head of the stud and a layer of reinforcement. The head of a 19mm stud is usually 10mm thick.

Beams 4.4

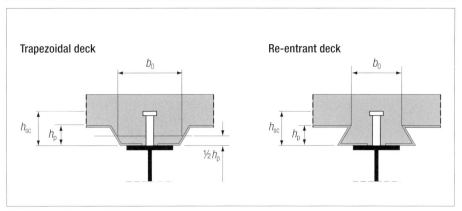

Fig 4.6 Beam with profile steel sheeting parallel to the beam

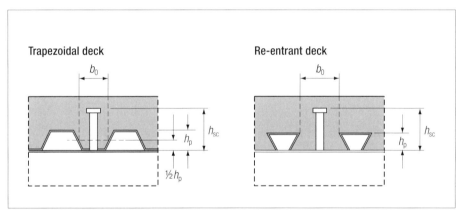

Fig 4.7 Beam with profile steel sheeting perpendicular to the beam

It is accepted in the UK that the depth of the decking in these expressions does not include small re-entrant stiffeners in the crest of the profile provided the width of the top crest is at least 110mm and the stiffener is not greater than 15mm in height nor 55mm in width. It should be noted that such stiffeners do need to be included when calculating the moment resistance of the beam.

4.4.7.3 Resistance of headed studs used with profiled steel sheeting to NCCI
Recent research has shown that the design resistances of 19mm diameter headed studs in trapezoidal deck transverse to the beam differs significantly in some cases from those calculated using the reduction factors in Section 4.4.7.2. This research has also taken account of the fact that it is not

4.4 Beams

normal practice in the UK to place reinforcement 30mm below the head of the stud. SCI note PN001a-GB[70] gives the design resistances but instead of modifying the partial safety factor it suggests different values for the reduction factor k_t.

The values given are based on the following conditions:
- The height of the trapezoidal decking is not less than 35mm nor greater than 80mm.
- The mean width of the troughs b_0 is not less than 100mm.
- The number of studs per trough is not more than 2.
- The as welded height of the studs is at least 95mm (the as welded height for through deck welded studs is assumed to be 5mm less than the nominal stud height).
- The ultimate strength of the studs should not be taken as greater than 450N/mm².
- The as welded height of the studs is at least 35mm greater than the height of the trapezoidal profile.
- The nominal thickness of the trapezoidal profile sheeting is not less than 0.9mm (bare metal thickness 0.86mm).
- Where there is a single stud per trough it should be placed in the central position. If this is not possible it may be placed in the favourable position such that the zone of concrete in compression in front of the stud is maximised.
- Where there are two studs per trough these should be placed in the central position. Where this is not possible they should be placed one either side of the centre of the trough.

Provided the above conditions apply k_t may be taken as the value from equation 6.23 from EC4-1-1 for one stud per trough. For two studs per trough it may be taken as the value from equation 6.23 multiplied by 0.7 if the reinforcement is above the heads of the studs and 0.9 if it is at least 10mm below the heads of the studs.

4.4.8 Spacing and minimum amount of shear connection

The longitudinal spacing of studs should be at least $5d$ and the transverse spacing should be at least $4d$. They should also be placed so that there is not less than 20mm between the edge of the stud and the edge of the flange.

Stud shear connectors may be spaced uniformly between critical sections of the beam. A number of studs should be provided so that a minimum amount of shear interaction is achieved. The degree of shear connection η is the ratio of the force in the slab from the shear connectors at the point of maximum moment $n_s P_{Rd}$ to the lesser of the axial resistance of the steel or concrete.

$$\eta = \frac{n_s P_{Rd}}{Min(R_c, N_{pl,a})}$$

The minimum degree of shear connection depends on the span L of the beam and the yield strength of the steel f_y.

EC4 Part 1-1[4] gives the following expressions for the minimum degree of interaction. They are based on the shear studs having a slip capacity of 6mm.

For $L \leqslant 25$m: $\eta \geqslant 1 - \left(\dfrac{355}{f_y}\right)(0.75 - 0.03L), \eta \geqslant 0.4$ (6.16 EC4-1-1)

For $L > 25$m: $\eta = 1$ (6.17 EC4-1-1)

Research has shown that studs with an overall height of not less than 95mm, and with a shank diameter of 19mm, which are through-deck welded into steel decking orientated perpendicular to the beam, have a slip capacity of over 10mm. As explained in SCI note PN002a-GB[71], this allows less conservative limits for beams not exceeding 25m in span. PN002a-GB gives the following expression:

For $L \leqslant 25$m: $\eta \geqslant 1 - \left(\dfrac{355}{f_y}\right)(1.433 - 0.054L), \eta \geqslant 0.4$

Figure 4.8 shows the above relationships. It should be noted that, if the expressions in Part 1-1 are used, the calculation of deflections is simplified if the amount of interaction is at least 0.5 (see Section 4.4.10).

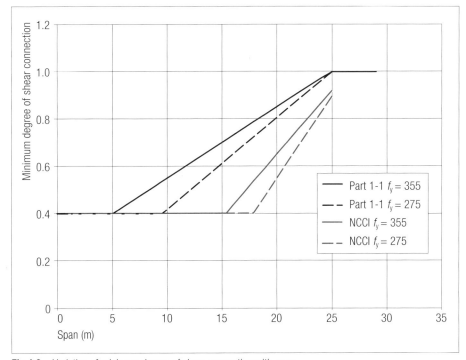

Fig 4.8 Variation of minimum degree of shear connection with span

4.4 Beams

4.4.9 Transverse shear

The section of slab either side of the beam needs to be checked to ensure that the shear force from the connectors can be transferred into the slab.

The force to be transferred is calculated from the shear resistance of the studs. This is divided either side of the beam in ratio to the effective widths of the concrete on the two sides. The longitudinal shear stress is calculated by taking this force over the depth of concrete above the decking. The percentage transverse reinforcement required to give a shear strength greater than the applied stress can be found using Figure 4.9. This Figure assumes the characteristic strength of the reinforcement is 500N/mm². The percentage reinforcement and the shear stress must be based on the depth of concrete above the deck h_c. Typical percentages for mesh sizes and typical depths above the deck are given in Table 4.10.

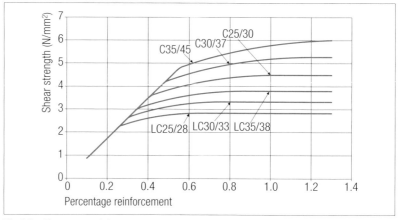

Fig 4.9 Transverse reinforcement

Table 4.10 Percentage reinforcement

Depth above deck h_c (mm)	Mesh size			
	A142	A193	A252	A393
60	0.24	0.32	0.42	0.66
70	0.20	0.28	0.36	0.56
80	*0.18*	0.24	0.32	0.49
90	*0.16*	0.21	0.28	0.44
100	*0.14*	*0.19*	0.25	0.39

Note
Values in italics are less than the minimum reinforcement required by EC4 (see this *Manual* Section 4.4.10).

Beams 4.4

Where the decking is transverse to the beam the decking can contribute and reduce the requirements for transverse shear reinforcement. Guidance is given in EC4[4] but conservatively the percentage reinforcement can be reduced by the following amount to allow for the strength of the connections of the deck to the beam at a butt joint:

$$\frac{P_{pb,Rd}\gamma_S}{sh_c f_{sk}}$$

$P_{pb,Rd}$ is the bearing resistance of the stud and expressions are given in EC4. It is the resistance of the connection of the steel deck to the top flange by means of the weld to the stud. A generally conservative value is 10kN. The stud spacing s to be used is the spacing of studs on each side of the beam. If there is only one stud per trough the studs will generally be staggered at a butt joint and s is twice the trough spacing.

Where the beam is adjacent to a longitudinal slab edge such that the distance from the edge of the slab to the nearest row of connectors is less than 300mm there are additional requirements for reinforcement. Transverse reinforcement should be supplied by horizontal U bars passing around the studs. The diameter of the U bars should be not less than half the stud diameter and they should be as low as possible in the slab.

4.4.10 Serviceability and deflections

Eurocode 4 Part 1-1[4] states that deflections due to loading applied to the composite member should be calculated using elastic analysis. For continuous up-propped beams it gives guidance on allowing for yielding of structural steel over supports. For the simply supported beams covered by this *Manual*, the calculated stress in the steel beam at serviceability limit state can, in some circumstances, also exceed yield but the code gives no guidance on how or whether to allow for this. For typical composite beams with UB sections the end connections will usually have some strength and stiffness that will reduce the actual stresses and deflections, and the amount and extent of any yielding that occurs will generally be small. The imposed load used to calculate the deflection is also likely to be higher than that which will normally be carried by the floor. Therefore it will not usually be necessary to check for yield of the steel at service load, elastic calculations of deflection will be sufficient. If the situation is not typical (e.g. unusual end conditions and/or the deflection calculation is critical) it may be necessary to limit the steel stress under service loads to a value less than the yield strength so that the deflection calculation is valid. For the arrangement shown in Figure 4.5 with a dead load equal to the live load and 75% of the dead load applied during construction the steel stress is likely to exceed the yield strength for beams where the composite resistance exceeds 1.5 times the non-composite resistance.

Deflections after construction are calculated assuming a composite second moment of area. For floors not subject to storage this can be calculated assuming the modulus of elasticity of the concrete is half the short term value given in

4.4 Beams

Tables 4.7 and 4.8. Figure 4.5 shows the increase in second moment of area for a composite beam compared to the steel section for typical slab dimensions and loading. Ratios are given for composite beams with normal weight (nwt) and lightweight (lwt) concrete. Provided the amount of interaction η is at least 0.5 and it complies with the expressions in EC4 Part 1-1[4], no account needs to be taken of the partial shear connection in the calculation of deflections. If η is less than 0.5 or the rules for the minimum degree of interaction in PN002a-GB[71] have been used the following expression can be used to calculate the increased deflection:

$$\delta = \delta_c + \alpha(\delta_a - \delta_c)(1 - \eta)$$

where:
α is 0.5 for propped construction and 0.3 for un-propped construction
δ_a is the deflection of the steel beam acting alone
δ_c is the deflection of the composite beam assuming $\eta=1$.

When normal weight concrete is used and the ratio of the span to overall beam depth is not greater than 20, shrinkage does not normally need to be taken into account. In other cases the additional deflection due to shrinkage can be calculated from the following expression:

$$\delta_s = \frac{e_s L^2}{8a\left(1 + \frac{\left(I_a + \frac{b_{eff}h_c^3}{12n}\right)\left(\frac{1}{A} + \frac{n}{b_{eff}h_c}\right)}{a^2}\right)} = \frac{e_s L^2 b_{eff} h_c z_{comp}}{8n I_{comp}}$$

The shrinkage strain e_s can be taken as 325×10^{-6} for normal weight concrete and 500×10^{-6} for lightweight concrete. I_a and I_{comp} are the second moments of area of the steel beam and composite section, A is the area of the steel beam, n is the modular ratio, z_{comp} the distance between the centroid of the concrete and composite section and a is the distance between the centroid of the concrete and the steel i.e. $a=0.5(h+h_c+2h_p)$.

It is possible to take the shrinkage deflection as $e_s L^2/(8a)$ but the correct value could be between 70 and 40% of this for typical beam sizes so this approach may be too conservative.

Shrinkage is time dependent. For typical slabs on metal deck, 100mm thick on average, approximately 50% will occur within the first 4 months and 75% within the first year after casting.

This *Manual* assumes the composite beams are designed as simply-supported. Where the slab is continuous over the support there is the risk of cracking. This is similar to the situation for floor slabs discussed in Section 3.4.3. EC4[4] requires a minimum percentage of reinforcement in the slab but this will not prevent cracking. The minimum is 0.2% for unpropped construction and 0.4% for propped construction. This is based on the area of concrete above the deck.

5 Columns in braced multi-storey buildings

5.1 Columns

This Section describes the design of uncased columns for braced multi-storey construction that are subject to compression and bending in simple construction.

In general the columns are designed as simple construction i.e. it is assumed that the moments are due only to nominal eccentricities (see Section 5.4d) and small moments due to the presence of cantilevers.

The resistance of a column is based on the cross sectional area of the member and its effective length. The resistance to axial loads is further reduced by the simultaneous application of bending moments, giving additional stresses in the extreme fibres of the column cross section. The resistance of columns with axial compression combined with bending moments may be checked by adding the component resistances in axial loading and bending moments in several locations along the column. Particular points occur where the column is likely to buckle and its axial resistance will be limited by its slenderness, and at positions of connections where bending moments are likely to be of maximum values.

Unless there are particular reasons to use a lower value the effective length of columns in braced multi-storey buildings should be taken as the storey height. Resistances for UC sections are given in Appendix B.

5.2 Column selection

Before selecting a trial section it is necessary to note that elements and cross-sections have to be classified as class 1, class 2, class 3 or class 4 in combined compression and bending according to the limiting width/thickness ratios stated in EC3 Part 1-1[1] clause 5.5.2 and Table 5.2. If the section is class 4 the compressive resistance must take account of a reduced effective area. In this *Manual* class 4 sections are not considered. It should be noted that all UCs and the universal beam sections shown in Table 5.1 are not class 4.

Other UB sections can be classified as not being class 4 if the axial compression is low enough that the web satisfies the limits for bending and compression in EC3 Part 1-1[1] Table 5.2 for the particular section. Using the lowest values for the range of UBs leads to the conclusion that if the ratio $N_{Ed}/N_{pl,Rd}$ is less than 0.51 for S275 material or 0.39 for S355 material the UB is not class 4.

Table 5.1　UB Sections which are not class 4 in direct compression

Grade S275		Grade S355
1016 × 305 × 487 [a,b]	356 × 171 × 57	1016 × 305 × 487 [a,b]
1016 × 305 × 437 [a,b]	305 × 165 × 54 [a]	1016 × 305 × 437 [b]
1016 × 305 × 393 [a,b]	305 × 127 × 48 [a]	610 × 305 × 238 [a]
914 × 419 × 388	305 × 127 × 42 [a]	533 × 312 × 272 [a,b]
610 × 305 × 238 [a]	305 × 127 × 37	533 × 312 × 219 [a,b]
610 × 305 × 179	254 × 146 × 43 [a]	533 × 312 × 182 [b]
533 × 312 × 272 [a,b]	254 × 146 × 37 [a]	533 × 210 × 138 [b]
533 × 312 × 219 [a,b]	254 × 146 × 31	457 × 191 × 161 [a,b]
533 × 312 × 182 [a,b]	254 × 102 × 28	457 × 191 × 133 [a,b]
533 × 312 × 150 [b]	254 × 102 × 25	457 × 191 × 106 [b]
533 × 210 × 138 [a,b]	203 × 133 × 30 [a]	406 × 178 × 85 [b]
533 × 210 × 122	203 × 133 × 25 [a]	305 × 165 × 54
457 × 191 × 161 [a,b]	203 × 102 × 23 [a]	305 × 127 × 48 [a]
457 × 191 × 133 [a,b]	178 × 102 × 19 [a]	305 × 127 × 42
457 × 191 × 106 [a,b]	152 × 89 × 16 [a]	254 × 146 × 43 [a]
457 × 191 × 98 [a]	127 × 76 × 13 [a]	203 × 133 × 30 [a]
457 × 191 × 89		203 × 133 × 25 [a]
457 × 152 × 82		203 × 102 × 23
406 × 178 × 85 [a,b]		178 × 102 × 19 [a]
406 × 178 × 74		152 × 89 × 16 [a]
356 × 171 × 67 [a]		127 × 76 × 13 [a]

Notes
a These sections are class 1 or 2 under compression.
b These sections are part of the Corus Advance range but are not in BS4[46].

5.3　Columns in simple construction

For simple multi-storey construction braced in both directions, according to UK practice[54], the columns should be designed by applying nominal moments only at the beam-to-column connections. The following conditions should be met:
(a) columns should be effectively continuous at their splices
(b) pattern loading may be ignored
(c) all beams framing into the columns are assumed to be fully loaded

(d) nominal moments are applied to the columns about the two axes
(e) the nominal moments are calculated using the full value of the imposed load, the axial load in the column can be calculated using a reduced imposed load to allow for the number of storeys supported as mentioned in Section 3.2
(f) nominal moments are proportioned between the length above and below the beam connection according to the stiffnesses I/L of each length, but where the ratio of I/L for the two storeys does not exceed 1.5 the moment can be divided equally above and below the beam connection
(g) nominal moments may be assumed to have no effects at the levels above and below the level at which they are applied.

Note The nominal moments mentioned in subclause (d) are the minimum moments to be used for column design. They are calculated as described in Section 5.4(d).

5.4 Design procedure

(a) Calculate the factored beam reactions. If the only variable action that needs to be considered is imposed load this is equal to 1.5 × imposed load + 1.35 × dead load from the beams bearing onto the column from each axis at the level considered. It may be necessary to calculate the reactions for different load factors for different load combinations.
(b) Calculate the factored axial force N_{Ed} on the column at the level being considered.
(c) Choose a section for the lowest column length. The following may be used as a guide to the size required:
203 UC for buildings up to 3 storeys high
254 UC for buildings up to 5 storeys high
305 UC for buildings up to 8 storeys high
356 UC for buildings from 8 to 12 storeys high
If UC sections are not acceptable choose a UB section that is not class 4, but note that as the slenderness ratio is less for a UB than a UC a larger area of section will be required.
(d) Calculate the nominal moments applied to the column about the two axes by multiplying the factored beam reactions by eccentricities based on the assumption that the loads act 100mm from the face of the column, or at the centre of a stiff bearing, whichever is greater.
If the beam is supported on a cap plate the load should be taken as acting at the edge of the column or edge of any packing.
(e) Obtain the nominal moments $M_{y,Ed}$ and $M_{z,Ed}$ applied to each length of the column above and below the beam connections by proportioning the total applied nominal moments, from Section 5.3 (d) according to the rule stated in Section 5.3 (e).

5.4 Columns in braced multi-storey buildings

(f) For I and H or RHS sections in simple construction where the only moments are those due to nominal eccentricity of connections, the Access Steel document SN048a[55] has shown that the following expression can be used to verify the member:

$$\frac{N_{Ed}}{\chi_{min}(Af_{yd})} + \frac{M_{y,Ed}}{\chi_{LT}(W_y f_{yd})} + \frac{1.5M_{z,Ed}}{W_z f_{yd}} \leqslant 1$$

where:
N_{Ed} is the design axial load on the member
$M_{y,Ed}$ is the design moment about the major axis
$M_{z,Ed}$ is the design moment about the minor axis
A is the cross section area of the member
W is the modulus of the section for the particular axis, it will be the plastic modulus for class 1 or 2 sections and the elastic modulus for class 3 sections
f_{yd} is the design yield strength of the material = f_y/γ_{M1}
f_y is the yield strength of the material
γ_{M1} is the partial material factor – given as 1.0 in the UK National Annex
χ_{min} the minimum of the reduction factors for flexural buckling determined as follows:
(i) Determine the effective length of the member about both axes
(ii) Calculate the slenderness ratio $\lambda = L_{eff}/i$ about the appropriate axis. L_{eff} is the effective length and for braced multi-storey buildings using simple construction it should be taken as the storey height of the column
(iii) Determine the non-dimensional slenderness

$\bar{\lambda} = \lambda/(93.9\varepsilon)$ where $\varepsilon = \sqrt{235/f_y}$

Note that for grade S275 steel, $93.9\varepsilon = 86.8$ and for S355 steel, $93.9\varepsilon = 76.4$
(iv) Select the column curve from Table 5.2 for the appropriate section, if the selected section is not in this table refer to EC3 Part 1-1[1]. For columns Figure 4.3 also applies and may be used to select the value of χ for the appropriate column curve.

Columns in braced multi-storey buildings 5.5

Table 5.2 Selection of buckling curve

Cross section	Limits		Buckling about axis	Buckling curve
Rolled I sections	$h/b > 1.2$	$t_f \leqslant 40$mm	y-y z-z	a b
		$40\text{mm} < t_f \leqslant 100\text{mm}$	y-y z-z	b c
	$h/b \leqslant 1.2$	$t_f \leqslant 100$mm	y-y z-z	b c
		$t_f > 100$mm	y-y z-z	d d

5.5 Columns with additional moments

If the column is subject to moments other than from beam eccentricity the following expressions can be used in place of that in Section 5.4[56].

For class I and H sections (susceptible to lateral torsional buckling):

$$\frac{N_{Ed}}{\chi_{min}(Af_{yd})} + \frac{M_{y,Ed}}{\chi_{LT}(W_y f_{yd})} + C_{mz}\frac{M_{z,Ed}}{W_z f_{yd}} \leqslant 0.78 \text{ for class 1 and 2, and 0.85 for class 3 and 4}$$

For RHS sections (not susceptible to lateral torsional buckling):

$$\frac{N_{Ed}}{\chi_{min}Af_{yd}} + C_{my}\frac{M_{y,Ed}}{W_y f_{yd}} + C_{mz}\frac{M_{z,Ed}}{W_z f_{yd}} \leqslant 0.85$$

The interaction limits of 0.85 and 0.78 are minimum values and apply at a particular axial load. The variation of the limits with applied load can be seen in Figure 5.1.

C_{my} and C_{mz} are uniform moment factors, they can conservatively be taken as 1.0. Smaller factors are given in Annex B to EC3 Part 1-1[1] and these are shown in Table 5.3. If uniform moment factors significantly less than 1.0 are used it will be necessary to check the local resistance of the section.

5.5 Columns in braced multi-storey buildings

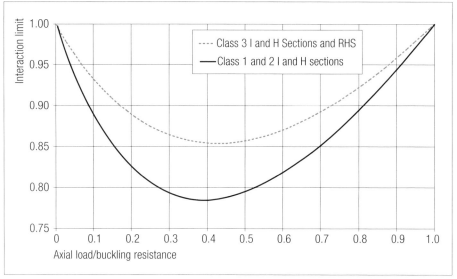

Fig 5.1 Variation of interaction limit with axial load

Table 5.3 Uniform moment factors

Moment diagram	Range	C_{my} and C_{mz}	
		Uniform loading	Concentrated load
M ⎯⎯ ψM	$-1 \leqslant \psi \leqslant 1$	$0.6 + 0.4\psi \geqslant 0.4$	
M_h M_s ψM_h	$0 \leqslant \alpha_s \leqslant 1$	$-1 \leqslant \psi \leqslant 1$	$0.2 + 0.8\alpha_s \geqslant 0.4$	$0.2 + 0.8\alpha_s \geqslant 0.4$
	$-1 \leqslant \alpha_s \leqslant 0$	$0 \leqslant \psi \leqslant 1$	$0.1 - 0.8\alpha_s \geqslant 0.4$	$-0.8\alpha_s \geqslant 0.4$
$\alpha_s = M_s/M_h$		$-1 \leqslant \psi \leqslant 0$	$0.1(1-\psi) - 0.8\alpha_s \geqslant 0.4$	$0.2(-\psi) - 0.8\alpha_s \geqslant 0.4$
M_h M_s ψM	$0 \leqslant \alpha_h \leqslant 1$	$-1 \leqslant \psi \leqslant 1$	$0.95 + 0.05\alpha_h$	$0.90 + 0.10\alpha_h$
	$-1 \leqslant \alpha_h \leqslant 0$	$0 \leqslant \psi \leqslant 1$	$0.95 + 0.05\alpha_h$	$0.90 + 0.10\alpha_h$
$\alpha_h = M_h/M_s$		$-1 \leqslant \psi \leqslant 0$	$0.95 + 0.05\alpha_h(1 + 2\psi)$	$0.90 + 0.10\alpha_h(1 + 2\psi)$

Notes
a For members with sway buckling mode the equivalent uniform moment factor should be taken as $C_{my} = 0.9$ or $C_{mz} = 0.9$ respectively
b C_{my} and C_{mz} shall be obtained according to the bending moment diagram between the relevant braced points as follows:

Moment factor	Bending axis	Points braced in direction
C_{my}	y-y	z-z
C_{mz}	z-z	y-y

6 Members in bracing systems

6.1 Introduction

This Section describes the design of bracing and other members that are subject to compression or tension only. It does not cover built up members; guidance on these is given in EC3 Part 1-1[1].

The compression resistance of members is based on the slenderness. Where this is high (e.g. the effective length L_{eff} divided by the radius of gyration about the relevant axis exceeds 180) care should be taken to make sure that all possible transverse loads on the member are allowed for including self weight.

6.2 Cross sectional areas

For the design of compression members the area taken should be the gross area and no allowance need be made for normal round holes filled with fasteners, although other holes, (e.g. oversized or slotted) should be taken into account. In the design of tension members the net area and gross area should always be considered. The gross area is the total area of the section reduced to allow for any openings greater than those required for fastenings. The net area is the gross area of the section reduced to allow for any holes for fasteners. When the holes for the bolts are staggered the area to be deducted for the bolts is taken as the maximum of:
– the area of the holes at the cross-section considered, (section a-a in Figure 6.1), and
– the area of holes in any diagonal less $\frac{s^2 t}{4p}$, (section b-b in Figure 6.1),

such that for the example shown below,

$$\text{area} = bt - \max\left(d_0 t \text{ or } t\left(2d_0 - \frac{s^2}{4p}\right)\right)$$

where:
t is the material thickness
d_0 is the hole diameter.

6.3 Members in bracing systems

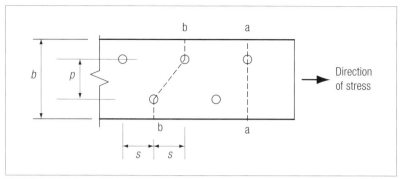

Fig 6.1 Net cross section

In the case of angles with holes staggered in both legs, p should be measured along the thickness of material as shown in Figure 6.2.

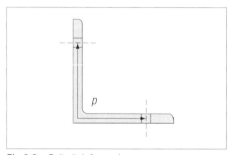

Fig 6.2 Bolt pitch for angles

6.3 Buckling lengths and slenderness ratio

Member buckling resistance is governed by the ratio of the relevant effective length L_{eff} divided by the radius of gyration i about the appropriate axis.

For web members in lattice construction the system length should be taken as the distance between the points of intersection of incoming members on any axis.

For chord members the system length should be taken as the distance between the intersection of web members in the plane of the truss and between the external restraints out of the plane of the truss.

The effective length L_{eff} should be taken as the system length unless a detailed analysis giving a smaller value is carried out by the engineer.

6.4 Resistance of sections in axial compression

6.4.1 General

The buckling resistance is critical for members subject to axial compression.

The design compressive force N_{Ed} should meet the following requirement:

$N_{Ed} < N_{b,Rd}$

$N_{b,Rd}$ is the design buckling resistance given by the expression:

$N_{b,Rd} = \chi A f_y / \gamma_{M1}$ (6.47 EC3-1-1)

where:
A is the gross area of the section. For class 4 sections (not covered by this Manual) a reduced effective area is used to allow for local buckling
χ is a reduction factor for buckling from Figure 4.3 for the maximum $\bar{\lambda}$. For hot-finished tubes curve a should be used, for angles curve b should be used and for plates, tees, channels and cold formed tubes curve c should be used
$\bar{\lambda}$ is the relative slenderness (see Section 1.5.3)
γ_{M1} is the partial factor for the material in buckling situations given a value of 1.0 in the UK National Annex.

If the element is fully restrained the local resistance can be used i.e. $A f_y / \gamma_{M0}$. This will only be different to the buckling resistance if the partial safety factors γ_{M0} and γ_{M1} are different; the UK National Annex gives them both the same value (1.0).

6.4.2 Angles in compression

For angles used as braces in compression, provided that the connection is made by at least two bolts and the supporting element provides appropriate end restraint, the eccentricities may be ignored and effective relative slenderness $\bar{\lambda}_{eff}$ obtained as follows (EC3 Part 1-1[1] BB.1.2):

for buckling about v-v axis $\bar{\lambda}_{eff,v} = 0.35 + 0.7 \bar{\lambda}_v$

for buckling about y-y axis $\bar{\lambda}_{eff,y} = 0.5 + 0.7 \bar{\lambda}_y$ (BB.1 EC3-1-1)

for buckling about z-z axis $\bar{\lambda}_{eff,z} = 0.5 + 0.7 \bar{\lambda}_z$

6.5 Members in bracing systems

Where the connection is made by one bolt the effective lengths must be taken as the system length and the eccentricity of the connection should be taken into account, see EC3.

$$\bar{\lambda} = \frac{L_{eff}}{i(93.9\varepsilon)}$$ about the relevant axis

where:
L_{eff} effective length about the relevant axis
i radius of gyration about the relevant axis
$\varepsilon = \sqrt{235/f_y}$

Note that for grade S275 steel, $\bar{\lambda} = \lambda/(86.8)$.

6.5 Resistance of sections in axial tension

6.5.1 General

For members subject to axial tension the design load N_{Ed} should meet the requirements of the following:
$N_{Ed} < N_{t,Rd}$ where $N_{t,Rd}$ is the tension resistance of the cross section
$N_{t,Rd}$ should be taken as the smaller of the following two expressions:

– The design plastic resistance of the cross-section:

$$N_{pl,Rd} = \frac{Af_y}{\gamma_{M0}}$$ (6.6 EC3-1-1)

where:
A is the gross area
f_y is the yield stress
γ_{M0} is a partial factor for the material, given a value of 1.0 in the UK National Annex.

– The design ultimate resistance of the section at the location of the fixing bolts:

$$N_{u,Rd} = 0.9\frac{A_{net}f_u}{\gamma_{M2}}$$ (6.7 EC3-1-1)

where:
A_{net} is the net area
f_u is the ultimate strength
γ_{M2} is a partial safety factor given as 1.1 in the UK National Annex to EC3 Part1-1.

Members in bracing systems 6.5

For connections using pre-loaded (e.g. HSFG) bolts designed to be slip resistant at the ultimate limit state (Category C connections), the plastic resistance of the net section at the holes for fasteners $N_{net,Rd}$ should be taken as not more than:

$$N_{net,Rd} = A_{net} f_y / \gamma_{M0} \qquad (6.8 \text{ EC3-1-1})$$

To ensure that the full ductile resistance is achieved before failure the plastic resistance $N_{pl,Rd}$ should be less than the ultimate resistance at the bolt holes $N_{u,Rd}$, i.e.

$$N_{u,Rd} > N_{pl,Rd}$$

This condition will be satisfied if:

$$0.9 \left\{ \frac{A_{net}}{A} \right\} > \left\{ \frac{f_y \gamma_{M2}}{f_u \gamma_{M0}} \right\}$$

6.5.2 Angles connected by one leg

EC3 Part 1-8[2] gives particular rules for angles connected by a single row of bolts in one leg. They may be treated as concentrically loaded and the eccentricity ignored provided that the ultimate resistance of the net section is determined as follows:

1 bolt: $\qquad N_{u,Rd} = \dfrac{2.0(e_2 - 0.5 d_0) t f_u}{\gamma_{M2}} \qquad$ (3.11 EC3-1-8)

2 bolts: $\qquad N_{u,Rd} = \dfrac{\beta_2 A_{net} f_u}{\gamma_{M2}} \qquad$ (3.12 EC3-1-8)

3 bolts: $\qquad N_{u,Rd} = \dfrac{\beta_3 A_{net} f_u}{\gamma_{M2}} \qquad$ (3.13 EC3-1-8)

where:
e_1, e_2, and p_1 are defined in Figure 6.3
A_{net} is the net area of angle
d_0 is the hole diameter
f_u is the ultimate tensile strength of the material
β_2 and β_3 are reduction factors dependent on pitch of bolts as shown in Table 6.1.

6.6 Members in bracing systems

Table 6.1 Reduction factors β_2 and β_3

Pitch p_1	< 2.5 d_0	< 5.0 d_0
2 bolts β_2	0.4	0.7
3 bolts β_3	0.5	0.7

Note For intermediate values of p_1, β can be determined by interpolation.

For an unequal-leg angle connected by its smaller leg A_{net} should be taken as equal to the net section area of an equivalent equal-leg angle of leg size equal to that of the smaller leg.

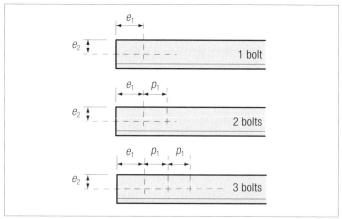

Fig 6.3 Edge and end distances

6.6 Bending and axial tension (EC3 Part 1-1 cL.6.2.9)

Where members are subject to combined bending and axial tension they can be checked by comparing the maximum stress to the yield strength of the material. Less conservative rules for certain elements can be found in EC3 Part 1-1[1].

Although it is not specifically mentioned in EC3, where the tension is not enough to prevent compression stresses and the section is susceptible to lateral torsional buckling the effect of buckling should be considered. A reasonable way to do this is to calculate a moment that gives the same compressive stress as the combined bending and tension. The member should then be checked for this moment according to Chapter 4.

7 Single-storey buildings – General

7.1 Introduction

This section offers advice on the general principles to be applied when preparing a scheme for a single-storey building. The basic principles are also applicable to mezzanine floors but the particular aspects of these floors are outside the scope of this *Manual* and reference should be made to BRE Digest 437[57].

7.2 Actions (loads)

7.2.1 General

Loading shall be based on the Eurocodes as described in Section 2.6. This should include equivalent horizontal loads from imperfections. Permanent and variable actions need to be defined. The BCSA has published a *Guide to Eurocode load combinations for steel structures*[58] that deals specifically with single-storey buildings and gives detailed guidance and examples.

7.2.2 Permanent actions G

Permanent actions are gravity loads and include the weight of the roof sheeting and equipment fixed to the roof, the structural steelwork, the ceiling and any services. The following approximate loads may be used for preliminary designs, in the absence of actual loads:

Roof sheeting and side cladding 0.1 to 0.2kN/m²
Steelwork 0.1 to 0.3kN/m²
Ceiling and services 0.1 to 0.3kN/m²

Although services loading is treated as a permanent action it is recommended that account is taken of the presence or not of this load by considering two different values of permanent load. Where the permanent load is unfavourable, the value should include the ceiling and services load. This is termed $G_{k,sup}$ in EC0[16]. Where the load is favourable, the ceiling and services load should be excluded ($G_{k,inf}$).

7.2 Single-storey buildings – General

7.2.3 Variable actions Q

Variable actions include live loads such as imposed loads, wind load and snow load. As mentioned in Section 2.8.1 imposed loads do not need to be applied with snow or wind loads.

Imposed loading is specified in the UK National Annex to BS EN 1991-1-1[12] as a minimum of 0.60kN/m^2 for pitched roofs of 30° to the horizontal or flatter. BS EN 1991-1-3[13] also gives the loads arising from the effects of uniformly distributed and drifting snow. According to the UK National Annex drifted snow on multi-span roofs, roofs closer to taller structures and at projections or obstructions are considered to be accidental loads. Wind loading varies with roof pitch and with the presence of dominant openings. Thermal loads should be considered but are not usually a problem if suitable movement joints are provided (see Section 2.5).

7.2.4 Equivalent horizontal forces

As described in Section 2.6.2, imperfections need to be allowed for and these are usually taken as equivalent horizontal forces (EHF). The values of imperfection in Table 2.1 can be used for single-storey buildings.

7.2.5 Load combinations

As mentioned in Section 2.8.2 the *Manual* only considers one of the equations in EC0[16] for load combinations and an alternative is available that, although more complex, could be more economic. The equation used leads to a number of combinations that need to be considered. These are given in Table 7.1 using the load factors from the UK National Annex to EC0[16]. Thermal loads are not included.

Table 7.1 Load combinations for single-storey buildings

Loadcase	Combination
Permanent + imposed load	$1.35G_k + 1.5Q_k$ + EHF
Permanent + snow load	$1.35G_k + 1.5S_k$ + EHF
Permanent + snow plus wind	$1.35G_k + 1.5S_k + 0.75W_k$ + EHF
Permanent + wind plus snow	$1.35G_k + 1.5W_k + 0.75S_k$ + EHF
Permanent + wind plus minimum vertical load	$1.0G_k + 1.5W_k$ + EHF
Permanent + accidental	$1.0G_k + 1.0A_d + 0.2W_k$ + EHF

As mentioned in Section 2.8.2 imposed loads on roofs do not need to be applied together with wind, or snow loads, hence the first combination does not include wind. The second combination needs to be considered if there are members for which the wind load is favourable. The first and second combination can be combined by using the maximum value of the imposed load and uniform snow load. The wind load W_k is probably more than one case, corresponding to wind from different directions.

7.3 Material selection

In the UK, material for rafters and columns should generally be of grade S275, although in many cases it may be cost-effective to use grade S355 unless deflection is likely to be critical. For example, S355 may be used for latticed and trussed members. Class 8.8 bolts should normally be used throughout and, in the interests of simplicity, a uniform diameter should be adopted for all bolted connections. M20 bolts are suitable for general use except for heavily loaded connections where M24 or larger bolts can be used. For light latticed frames M16 bolts may be appropriate. In order to save confusion and mistakes on site it is recommended that a single size and type of bolt be used on a project, as far as is practical.

7.4 Structural form and framing

The most common forms of single-storey frames are:
- portal frames with pinned bases
- posts with pinned bases and pitched trusses
- posts with pinned bases and nominally parallel lattice girders.

Fixed bases can be used instead of pinned bases, but these generally require larger and more costly foundations which may not justify any saving in the weight of steel frame resulting from fixing column bases.

The design should be based on:
- The spacings between frames and spans given in Table 7.2, which are likely to be economic.
- Providing longitudinal stability of portal frames against horizontal forces by placing vertical bracing in the side walls, deployed symmetrically, wherever possible.
- Providing stability of post-and-truss or post-and lattice girder frames against lateral forces in two directions, approximately at right angles to each other, by arranging suitable braced bays deployed symmetrically wherever possible.
- Providing bracing in the roof plane of all single-storey construction to transfer horizontal loads to the vertical bracing.
- Providing bracing to the bottom of members of trusses or lattice girders, if needed, to cater for reversal of forces in these members because of wind uplift.
- The provision of movement joints for buildings in the UK whose plan dimension exceeds 50m.
- Purlins, where possible, having support at node points for lattice girders and trusses. Alternatively, the effects of local bending at the lattice members needs to be taken into account.

- The arrangement of structural framing taking account of openings for doors and windows, support for services and foundations, (e.g. columns immediately adjacent to site boundaries may require special foundations).
- The provision of framing at openings to transfer horizontal forces to braced elements.
- Consideration of the use of shorter end bays as that may allow a more efficient design.

Table 7.2   Typical spacings and spans for single-storey buildings

Type	Spacing (m)	Span (m)
Portals	5.0 – 7.0	up to 60[59]
Post-and-pitched truss	4.5 – 7.5	18 – 25
Post-and-lattice girder	4.5 – 7.5	20 – 40

The choice of structural form will depend on such factors as the appearance of the building, the extent that supports are required in the roof for ceiling and services and on overall economy.

7.5   Fire resistance

Fire protection should be considered for those frames that provide lateral stability to perimeter or party walls and which are required to have a fire rating. Protection can be achieved by casing either the whole frame or only the columns in fire resistant material. EC3 Part 1-2[25] may be used to determine the fire resistance of steel structures. If only the columns are so treated, the stability of the columns acting as cantilevers without the lateral support provided by the rafters or trusses must be checked to ensure the stability of the whole structure is not in danger, if the effectiveness of these members is removed by the action of a fire. Further guidance on this is given in SCI publication P313, *Single storey steel framed buildings in fire boundary conditions*[60]. Fire protection should also be considered for structural members supporting mezzanine floors enclosed by single-storey buildings.

7.6   Corrosion protection

Structural steelwork should be protected from corrosion. See Section 3.6 of this *Manual*.

7.7 Bracing

Choose the location and form of bracing in accordance with the recommendations in Sections 7.4 and 2.3.3. Typical alternative locations are shown on Figure 7.1 for single-storey buildings.

Wind loads on the structure should be assessed for the appropriate load combinations and divided into the number of bracing bays resisting the horizontal forces in each direction. The distribution of the horizontal forces between the bracing bays should be made on a rational basis taking account of the position and likely stiffness of each bracing bay.

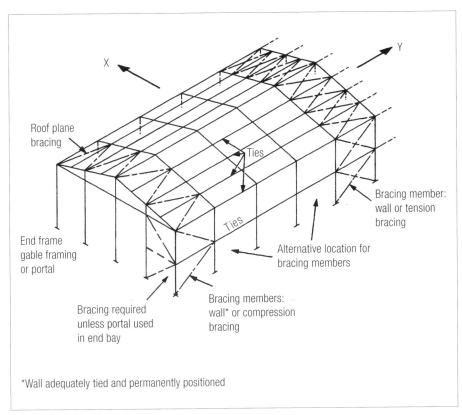

*Wall adequately tied and permanently positioned

Fig 7.1 Typical bracing systems

7.8 Purlins and side rails

It is normal practice to use cold-formed sections as purlins and side-rails. The resistance of these sections is provided by the manufacturer. They should be used in accordance with the manufacturers recommendations e.g. provision of anti-sag rods. It should also be confirmed that, where necessary, the cladding provides the restraint assumed by the manufacturer.

7.9 Roof and cladding materials

Although this *Manual* is concerned with the design of structural steelwork, it is essential at the start of the design to consider the details of the roof and cladding systems to be used, since these have a significant effect on the steelwork.

The choice of cladding material depends on whether the roof is flat or pitched. For the purposes of this *Manual* a roof will be considered flat if the roof pitch is less than 6°. It should be noted, however, that roofs with pitches between 6° and 10° will often require special laps and seals to avoid problems with wind-driven rain etc.

The variety of materials available for pitched roofs is vast and a complete presentation is beyond the scope of this *Manual*. However, a brief description of the most common cladding systems is included in Table 7.3. It summarises the salient features of the various types of lightweight roofing systems commonly used in the UK.

Profiled decking systems described in Table 7.3 for pitched roofs are often suitable for use as wall cladding. Where insulation is required it can be provided either as bonded to the sheeting or in a 'dry-lining' form with the internal lining fixed to the inside face of the sheeting rails. Fire protection of the walls of industrial buildings can be achieved by using boarding with fire-resistant properties on the inside face of the sheeting rails.

It is not uncommon to provide brickwork as cladding in industrial buildings for the lower 2.0–2.5m of industrial buildings, because of the vulnerability of less robust materials to mechanical damage. Where this detail is required it is usually necessary to provide a horizontal steel member at the top of the wall spanning between columns to support such brick panel walls against lateral loading.

Single-storey buildings – General   7.9

Table 7.3 Lightweight roofing systems and their relative merits

Description	Minimum pitch	Typical depth (mm)	Typical span (mm)	Degree of lateral restraint to supports	Comments
Galvanized corrugated steel sheets	10°	75 sinusoidal profile	1800 - 2500	Good if fixed direct to purlins	Low-budget industrial and agricultural buildings; limited design life, not normally used with insulated liner system
Fibre-cement sheeting	10°	25 - 88	925 - 1800	Fair	Low-budget industrial and agricultural buildings; brittle construction usually fixed to purlins with hook bolts
Profiled aluminium sheeting (insulated or uninsulated)	6°	30 - 65	1200 - 3500	Good if fixed direct to purlins	Good corrosion resistance but check fire requirements and bi-metallic corrosion with mild-steel supporting members
Profiled coated steel sheeting (insulated or uninsulated)	6°	25 - 65	1500 - 4500	Good if fixed direct to purlins	The most popular form of lightweight roof cladding; used for industrial type buildings; wide range of manufacturers, profile types and finishes
'Standing seam' roof sheeting (steel or aluminium)	2°	45	1100 - 2200	No restraint afforded by cladding, clip fixings	Used for low-pitch roof and has few or no laps in direction of fall; usually requires secondary supports or decking, which may restrain main purlins
Galvanized steel or aluminium decking systems	Nominally flat	32 - 100	1700 - 6000	Very good	Used for flat roof with insulation, vapour barrier and waterproof membrane over; fire and bimetallic corrosion to be checked if aluminium deck used
Timber	Nominally flat	General guidance as for timber floors			
Reinforced woodwool slabs	Nominally flat	50 - 150	2200 - 5800	Good if positively fixed to beam flanges	Pre-screeded type can reduce the risk of woodwool becoming saturated during construction. Wetting, through condensation or leaks, can significantly reduce the strength

8 Portal frames with pinned bases

8.1 Introduction

Plastic methods are commonly used for the design of portal frames, resulting in relatively slender structures. In-plane and out-of plane stability of both the frame as a whole and the individual members must be considered.

The *Manual* gives methods for initial sizing of elements and equations so that in-plane frame stability can be checked by hand but these must only be used in those cases where gravity loading is critical. Because wind loading often governs the design of portal frames, it is preferable to use specialist software, although the methods herein may be suitable for preliminary design.

Elastic frame analysis may be used to obtain the forces and moments on the frame (actions). It will also be necessary where deflections are critical. The members should then be designed using the procedures given in Sections 4 and 5 of this *Manual* for the design of beams and columns. In addition, sway and snap-through stability checks should be carried out as for plastic design. Base stiffness is important for the design of these frames and guidance is given in an Access Steel document[61].

For multibay frames of equal spans, where the same rafter section size is used throughout, the design is almost invariably governed by the external bays. The internal columns are subjected to very little bending unless the loading is asymmetrical and may therefore conservatively have the same size as the columns for the single-span situation. Because internal columns will not usually be provided with the same restraints as external columns, their stability should be checked considering the effective restraints actually provided.

Eaves deflections of pitched multi-bay frames should be carefully checked, as the horizontal deflections will be cumulative. This applies particularly to frames which have a steep pitch.

8.2 Plastic analysis

Guidance is given on the design of single-span portals with pinned bases and where wind loading does not control the design. The procedure for the plastic method of design is given in the following sequence:
(1) sizing of rafters and columns
(2) check on sway and snap-through stability
(3) check on serviceability – deflection
(4) check on position of plastic hinges and calculation of resistance of frame
(5) checks on stability of plastic hinges, rafter, haunch and columns.

8.3 Sizing of rafters and columns

The initial trial sizing is carried out by the selection of members from graphs. Only gravity loads are considered and horizontal loads (e.g. wind) are ignored. This method is based on the following assumptions:
- Plastic hinges are formed at the bottom of the haunch in the column and near the apex in the rafter, the exact position being determined by the frame geometry.
- The depth of the haunch below the rafter is approximately the same as the depth of the rafter.
- The haunch length is 10% of the span of the frame, an amount generally regarded as providing a balance between economy and stability.
- The moment in the rafter at the top of the haunch is $0.87M_p$, and it is assumed that the haunch region remains elastic.
- The calculated values of M_p are provided exactly by the sections and that there are no stability problems. Clearly these conditions will not always be met and the chosen sections should be fully checked for all aspects.

8.4 Design procedure

The various frame dimensions are shown on Figure 8.1. This is not a rigorous design method, it is a set of rules to arrive at initial member sizes.

8.4 Portal frames with pinned bases

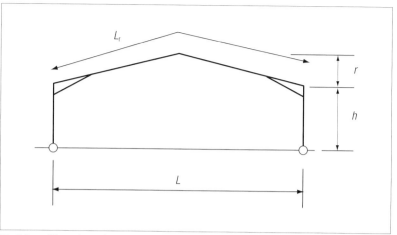

Fig 8.1   Dimensions of portal

(1) Calculate the span/height to eaves ratio = L/h
(2) Calculate the rise/span ratio = r/L
(3) Calculate the total design load FL on the frame (see Section 7.2) and then calculate FL^2, where F is the load per unit length on plan of span L (e.g. $F = qs$, where q is the total factored load per m² and s is the bay spacing)
(4) From Figure 8.2 obtain the horizontal force ratio H_{FR} at the base from r/L and L/h
(5) Calculate the horizontal force at base of span $H = H_{FR} FL$
(6) From Figure 8.3 obtain the rafter M_p ratio M_{pr} from r/L and L/h
(7) Calculate the M_p required in the rafter from M_p (rafter) = $M_{pr}FL^2$
(8) From Figure 8.4 obtain the column M_p ratio M_{pl} from r/L and r/h
(9) Calculate the M_p required in the column from M_p (column) = $M_{pl}FL^2$
(10) Determine the plastic moduli for the rafter $W_{pl,y,R}$ and column $W_{pl,y,C}$ from:

$$W_{pl,y,R} = M_p \text{ (rafter)}/f_y \quad \text{and} \quad W_{pl,y,C} = M_p \text{ (column)}/f_y$$

where f_y is the yield strength obtained from Table 2.6.

Using these plastic moduli, the rafter and column sections may be chosen from the range of sections classified as plastic and given in publications e.g. SCI P363[6].

Portal frames with pinned bases 8.4

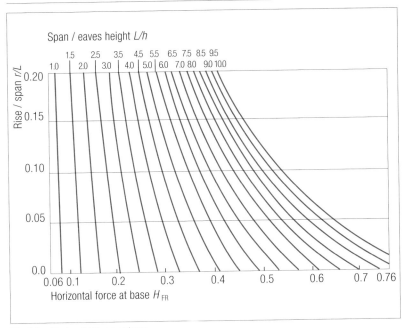

Fig 8.2 Horizontal force at base

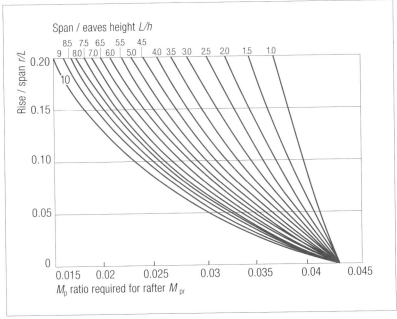

Fig 8.3 M_p ratio required for rafter M_{pr}

8.5 Portal frames with pinned bases

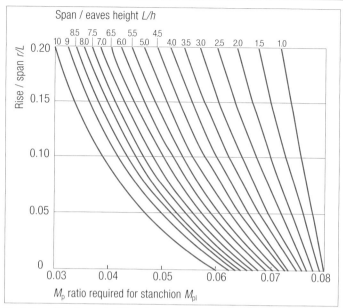

Fig 8.4 M_p ratio required for column M_{pl}

8.5 Sway and snap-through stability

8.5.1 General

Two modes of stability failure have been identified, the first may occur in any frame and is called 'sway stability'. The mode of failure is caused by the change in frame geometry arising from applied loading which gives rise to the PΔ effect, when axial loads on compression members displaced from their normal positions give moments which reduce the frame's resistance.

The second mode can take place when the rafters of three or more bays have their sections reduced because full advantage has been taken of the fixity provided in the valleys. In this case the risk of snap-through should be considered.

If either of the stability checks is not satisfied or vertical loading is not the critical case, or if one of the other restrictions given in Section 8.3 is not satisfied, other methods must be used. These are outside the scope of this *Manual*.

8.5.2 Sway stability check

This approximate method for the check of sway stability may be useful for establishing initial trial sizes, there is no equivalent in EC3.

Portal frames with pinned bases 8.5

The method given below may be used only if the frame is not subject to loads from valley beams or crane gantries or other concentrated loads larger than those from purlins. In addition, each bay should satisfy the following conditions:
– The rafter is symmetric about the apex.
– The span L does not exceed 5 times the column height h.
– The height h_r of the apex above the tops of the columns does not exceed 0.25 times the span L.

The check is based on the condition in the following formula:

$$\frac{L_b}{h_b} \leqslant \frac{44}{\Omega} \frac{L}{h_c} \frac{\rho}{(4+\rho L_r/L)} \frac{275}{f_y}$$

where:

$$L_b = L - \left[\frac{2h_h}{(h_s+h_h)}\right] L_{hp}$$

$$\rho = \frac{2I_{y,c}}{I_{y,r}} \frac{L}{h_c} \text{ for single bay frames or } \frac{I_{y,c}}{I_{y,r}} \frac{L}{h_c} \text{ for multibay frames}$$

L = span of the bay
L_{hp} = horizontal projected length of haunch, see Figure 8.5. If the haunches at each side of the bay are different the mean value should be taken
h_b = minimum depth of rafters, see Figure 8.5
h_h = additional depth of the haunch, see Figure 8.5
h_s = depth of rafter allowing for its slope, see Figure 8.5
h_c = column height
$I_{y,c}$ = second moment of area of column, =0 if not rigidly connected to the rafter or if the rafter is supported on a valley beam
$I_{y,r}$ = second moment of area of rafter at its shallowest point
f_y = design strength
L_r = total developed length of the rafter, see Figure 8.1
Ω = F_r/F_0, the ratio of the arching effect of the frame
where:
F_r = factored total vertical load on the rafter
F_0 = maximum uniformly distributed load for plastic failure of the rafter treated as a fixed end beam of span L where $F_0 = 16W_{pl,y,R}f_y/L$.

If the condition given in the formula is satisfied then the frame should remain stable under gravity loads and deflections under this load should not seriously affect strength. If it is not satisfied then the frame member sizes can be adjusted until the L_b/h_b limit is satisfied. Alternatively more elaborate methods can be used to assess the frame. These will usually involve specialist software although Access Steel document SN033[62] shows how normal analyses can be used to assess second order effects.

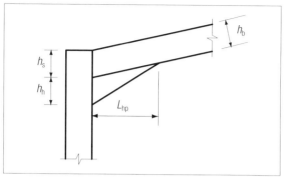

Fig 8.5 Dimensions of a haunch

8.5.3 Snap-through stability check

This should be carried out for frames of three or more spans, as in each internal bay snap-through may occur because of the spread of the columns and inversion of the rafter. The Access Steel document SN033[62] gives an expression for the elastic critical load factor for snap-through and to ensure this is not less than 10 the rafter slenderness should be such that:

$$\frac{L_b}{h_b} < \frac{22(4 + L/h_c)}{4(\Omega - 1)} \left(1 + \frac{I_{y,c}}{I_{y,r}}\right) \frac{275}{f_y} \tan 2\theta$$

where θ is the rafter slope.

No limit need be placed on L_b/h_b when $\Omega < 1$.

8.6 Serviceability check – deflection

Deflections under service loading can govern the design of portal frames. Generally, codes do not give specific limits for portal frame deflections. The responsibility for selecting suitable limits rests with the designer. Deflections should not impair the strength or efficiency of the structure or its components, nor cause damage to cladding and finishes. Access Steel document SN035[63] does suggest limits for various conditions for some European countries including the UK.

It is recommended that deflections due to service loading should be calculated by computer analysis. The stiffness of a nominally pinned base may be taken as 20% of the column stiffness for the purpose of calculating deflections. In this case the base and foundations need to be designed for the resulting moment.

8.7 Check on position of plastic hinge in rafter and calculation of load resistance

To check that the correct mode of failure has been assumed a reactant diagram should be drawn. This is obtained by plotting the moments due to the applied forces and known moments at hinge locations, including bases. If the moments at all points in the frame are less than the values of M_p and equal to M_p only at the hinge locations then the assumptions may be considered as satisfactory. If M_p of the frame is exceeded at any point in the frame then the diagram must be adjusted to take this into account.

To check the position of the plastic hinge and the load resistance of the frame the following simple procedure may be carried out:
(1) consider a pinned based portal frame subject to vertical loading as shown in Figure 8.6
(2) calculate $H = M_p \text{(column)}/h_1$
(3) calculate r/L and L/h and then determine x from Figure 8.7
(4) take moments about the rafter hinge position giving:
$M \text{(rafter)} = w'Lx/2 - H h_2 - w' x^2/2$
(5) As the moment of resistance of the rafter is known (M_p) then the value of w' may be determined
(6) redesign the frame if the load resistance w' is less than the design load on the frame.

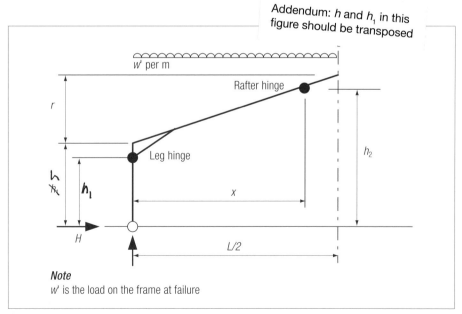

Addendum: h and h_1 in this figure should be transposed

Note
w' is the load on the frame at failure

Fig 8.6 Vertically loaded pinned-base portal frame

8.8 Portal frames with pinned bases

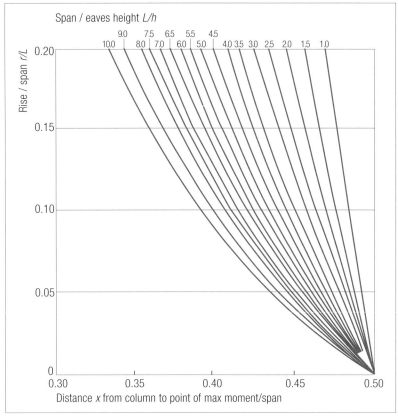

Fig 8.7 Distance x from column to point of maximum moment/span

8.8 Stability checks

The following stability checks should be carried out:
- restraint of plastic hinges
- stability of rafter
- stability of haunch
- stability of column.

8.8.1 Restraint of plastic hinges

A restraint should be provided to both flanges at each plastic hinge location. If this is not practicable, the restraint should be provided within a distance of half the depth of the member along the flanges of the member from the location of the plastic hinges.

Portal frames with pinned bases 8.8

The maximum distance from the hinge restraint to the next adjacent restraint should not exceed L_m (see Figure 8.8) where:

$$L_m = \frac{38 i_z}{\sqrt{\frac{1}{57.4}\left(\frac{N_{Ed}}{A}\right) + \frac{1}{756 C_1^2}\left(\frac{W_{pl,y}^2}{A I_t}\right)\left(\frac{f_y}{235}\right)^2}}$$
(BB.5 EC3-1-1)

where:
N_{Ed} is the design value of the compression force in the member

$\dfrac{W_{pl,y}^2}{A I_t}$ for the haunch is the maximum value in the segment

A is the cross section area of the member at the location where the above ratio is a maximum
$W_{pl,y}$ is the plastic section modulus of the member
i_z is the minimum radius of gyration of the member
f_y is the yield strength (N/mm²)
C_1 is the uniform moment factor (see Section 4).

If the adjacent restraint is only on the tension flange then the maximum distance to the nearest restraint to both flanges should not be greater than the distance L_k (see Figure 8.8). This may be taken as L_s calculated as for the stability of haunch but with a taper factor of 1.0 (see Section 8.8.3).

The above expression can be used for columns and haunched rafters. If the rafter is tapered equation BB.10 from EC3 Part 1-1[1] must be used.

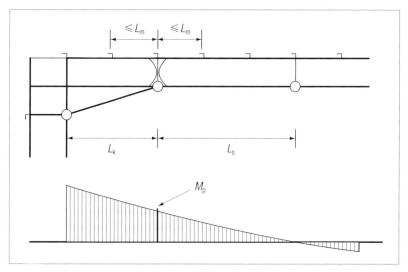

Fig 8.8 Restraint spacing

8.8 Portal frames with pinned bases

8.8.2 Rafter stability

Rafter stability should be checked for all load cases. Unless there is wind uplift, the following checks should be made:
- at the plastic hinge location near the ridge both flanges should be laterally restrained in order to provide torsional restraint
- a purlin or other restraint is needed on the compression flange at a distance determined in the same way as mentioned above for plastic hinge restraint
- further restraints to the top flange are required so that the rafter satisfies the requirements of Section 4 for beams without full lateral restraint
- in regions where there is compression on the bottom flange the procedure given for haunches in Section 8.8.3 should be applied using constants applicable to haunch/depth of rafter = 1.

8.8.3 Stability of haunch

Provided that the tension flange of the haunch is restrained, the maximum length between restraints to the compression flange of the haunch should be limited to L_s (see Figure 8.8) obtained from the equation that follows, provided that:
- the rafter is a UB section
- the haunch flange is not smaller than the rafter flange
- the depth of the haunch h_h is not greater than twice the depth of the rafter allowing for its slope h_s
- adjacent to plastic hinge locations the spacing of the tension flange restraint complies with Section 8.8.1
- the buckling resistance is satisfactory if it is checked as though it were a compression flange in accordance with Sections 4 or 5 using an effective length L_E equal to the spacing of the tension flange restraints.

L_s may conservatively be taken as:

$$L_s = \frac{\left(5.4 + \frac{600 f_y}{E}\right)\left(\frac{h}{t_f}\right) i_z}{c\sqrt{\left(5.4\left(\frac{f_y}{E}\right)\left(\frac{h}{t_f}\right)^2 - 1\right)}}$$

where:
i_z is the minimum radius of gyration of the rafter section
h is the depth of the rafter section
t_f is the flange thickness of the rafter section
c is the taper factor.

EC3 Part 1-1[1] gives methods to account for the shape of the moment diagram on the length L_s.

Portal frames with pinned bases   8.8

The taper factor c is 1.0 for a uniform member and for a haunched member is given by:

$$c = 1 + \frac{3}{\left(\frac{h}{t_f} - 9\right)} \left(\frac{h_h}{h_s}\right)^{2/3} \sqrt{\frac{L_h}{L_y}} \qquad \text{(BB.17 EC3-1-1)}$$

where:
h_h is the additional depth of the haunch, see Figure 8.9
h_s is the vertical depth of the un-haunched section, see Figure 8.9
L_h is the length of the haunch within the length L_y, see Figure 8.9
L_y is the length between points at which the compression flange is laterally restrained.

If the rafter is tapered the equation BB.16 in EC3 Part 1-1[1] must be used for the taper factor.

If no intermediate restraint is provided to the tension flange then the limiting length L_m to the nearest restraint on the compression flange should be calculated as for restraint of plastic hinges.

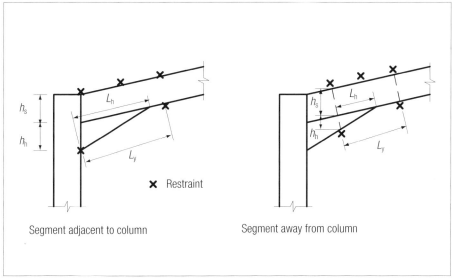

Fig 8.9   Dimensions defining taper factor

8.8 Portal frames with pinned bases

8.8.4 Stability of column

Near the top of the column a restraint should be provided at the location of the plastic hinge (underside of haunch), together with a further restraint at a distance below the position of the hinge restraint as described in Section 8.8.1. If the second restraint is only to the tension flange the maximum distance to a compression flange restraint should be L_s calculated using the expression in Section 8.8.3 with a taper factor of 1.0. The column should then be checked using the overall buckling check in Section 5. If further minor axis restraint is required it should be provided using side rails and stays. In the overall check, buckling about the major axis should also be considered using an effective length equal to the column height.

9 Lattice girders and trusses with pin-base columns

9.1 Lattice girders and trusses

Building with trussed roof members has now largely been superseded by the use of portal frames. However, there are still instances where trusses should be used. These include buildings with spans in excess of 50-60m, where portal frames become uneconomical, or because there are special crane gantry runways, etc. that require connections at the eaves level throughout the building. Sometimes these are also used purely for aesthetic reasons or because of the requirement of the surrounding buildings.

Trusses are generally split into two types.

The first type is the steep-pitched truss where the pitch is dependent on the type of roof covering. This pitch is usually 17½° for fibre cement sheeting, 22½° for tiled roofs and 30° for traditional slate roofs. This type of truss is now very rarely used.

The second is the type of truss that is basically a lattice girder with the top boom sloped at a pitch of 6° minimum and is used with the type of metal cladding preferred today. It is quite common for these trusses to be fabricated out of hollow section, but as hollow section is generally more expensive, fabrication and maintenance costs should be taken into account. Hollow sections are usually used where aesthetic considerations are paramount. The overall depth of this type of truss should be approximately 1/12 to 1/15th of the span. If the depth is less than this ratio deflection can be critical.

When using trusses with pin bases the horizontal forces can be resisted in one of two ways.

The first is by making the building a braced box, as shown in Figure 9.1 and transferring wind forces by the use of horizontal bracing to vertical bracing in the side and end gable. This may not be economic for large buildings.

Note that in the steep-pitch trusses this bracing should be at bottom tie level, but in the lattice girders it can be at top boom level. Additional bracing may be required in the plane of the top chords of steep-pitched trusses to restrain the roof trusses from lateral movement. It also may be required at the bottom chord level of lattice trusses if the bottom chord goes into compression e.g. due to wind load.

9.1 Lattice girders and trusses with pin-base columns

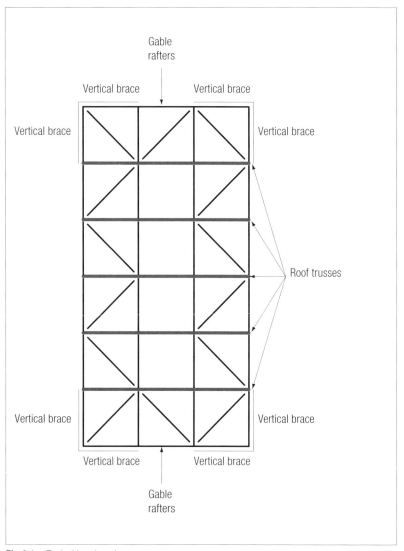

Fig 9.1 Typical bracing plan

In the second, the horizontal forces can be resisted as a moment at the eaves. In the case of a steep-pitch truss this involves making the eaves into a knee-braced truss (shown in Figure 9.2). For these structures horizontal deflection will often govern.

9.1 Lattice girders and trusses with pin-base columns

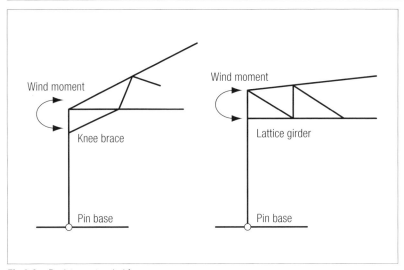

Fig 9.2 Resistance to wind forces

There is an alternative to both these methods which imposes moments on the foundation and this Section deals with pin bases only.

Lattice girders or trusses should be designed using the following criteria:
(1) Connections between internal and chord members may be assumed pinned for calculation of axial forces in the members.
(2) Members meeting at a node should be arranged so that their centroidal axis/or line of bolt groups coincide. When this is not possible the members should be designed to resist the resulting bending moments caused by the eccentricities of connections in addition to the axial forces.
(3) The joints will have some fixity and will attract secondary moments. In some cases the flexibility of the joint at the ultimate limit state will mean that these moments can be neglected. EC3 Part 1-8[2] provides particular guidance for members with hollow section joints designed according to Section 7 of that part. In those cases the secondary moments can be neglected provided that the ratio of the length of members to their depth in the plane of the truss is not less than 6.
(4) Where transverse loads are not applied at the node points the resulting moments must be taken into account. Where the secondary moments described in (3) can be neglected, web members do not need to be designed for any moment and the chords can be considered as continuous elements provided with simple supports at the panel points.
(5) The effective length of a chord member may be taken as the distance between connections of the web members in the plane of the girder or truss and the distance between the longitudinal ties or purlins in the

plane of the roof cladding. EC3 Part 1-1[1] allows lower values to be used in certain situations:
- For I and H section chords it may be taken as 0.9 times the system length for buckling in the plane of the truss.
- For hollow section trusses where the diameter of the brace is less than 0.6 times the chord diameter and there is no cropping or flattening of the brace at the end, it may be taken as 0.75 times the system length.
- For angle web members see Section 6 of this *Manual*.
- For web members of other sections it may generally be taken as 0.9 times the system length for buckling in the plane of the truss.

(6) Ties to the chords should be properly connected to an adequate restraint system.
(7) Bottom members should be checked for load reversal arising from uplift.
(8) Where secondary stresses arising from local bending exist, as a rule of thumb, a bending moment can be taken as 70% of the bending moment caused by a point load acting between points of support.

9.2 Determination of section sizes

Section sizes can then be determined as follows:

Compression members
Using the methods outlined in Section 6 of this *Manual* or by reference to published tables[6] giving the compression resistance.

Tension members
Using the methods outlined in Section 6 of this *Manual* or by reference to published tables[6].

Deflection
The deflection of the truss or lattice girder should be checked to see that serviceability with particular reference to roof drainage is not impaired. It will usually be found that deflection will not be a problem providing the aspect ratios of span/truss depth is approximately between 12 and 15. Preset cambers can be built into the girders during fabrication to offset the effects of dead load deflections.

Where splices occur in chord and internal members, careful consideration should be given to the effects of possible bolt slip in the splice connections.

9.3 Columns for single-storey buildings braced in both directions

Columns will generally need to be designed for a combination of axial force and moment in accordance with the procedures given in Section 5 of this *Manual*. The combinations given in Table 7.1 will need to be considered. Each combination should be considered separately but for initial sizing the maximum axial force and maximum moment from all the combinations can be used to give a conservative design.

Columns forming part of the braced bays will have different forces and moments and will need to be considered separately.

The effective length of the columns is usually taken as being equal to the actual length of the member (this is the system length).

Axial force
Axial force in columns will normally be the reaction from the roof truss plus the load from side cladding attached to the frame. The beneficial effect of wind suction on the roof can be neglected to reduce the number of load combinations to be considered. In the braced bays the column forces due to overall loads on the building will need to be calculated. The horizontal forces are those due to wind on the side walls, the horizontal component of wind on the roof, drag (if relevant) and the equivalent horizontal forces from the initial imperfections (see Section 2.6.2). Connections to foundations and the foundations themselves will need to be checked for any net tension in the columns due to wind suction and/or for the columns in braced bays.

Moments
There are three sources of moment on the columns and these are:
– The columns in the side walls will span between the base and eaves under the wind load.
– It is necessary to allow for a moment due to eccentricity of the vertical load applied to the columns from the truss. If the truss is supported on a cap plate the load is assumed to be applied at the face of the column or edge of packing, in all other cases the eccentricity should be taken as half the column depth + 100mm or to the centre of bearing whichever is greater.
– For the columns in the braced bays, there can be moments due to eccentricity of the bracing connections.

9.4 Columns for single-storey buildings braced in one direction only in the side walls and/or in the valleys

These columns are designed using the same procedures as set out in Section 9.3 with the following exceptions:
- The effective length of the columns in the plane of the truss will be longer than the length of the column.
- If the bases are pinned it should be at least twice the actual length. The moments, for the design of the columns forming part of the moment frame, will need to be established from elastic analysis of the frame under the various load combinations listed in Table 7.1.

10 Connections

10.1 Introduction

The connections for frames analysed with simple joints and single-storey portal frames with continuous joints are covered in this *Manual*.

The relationship between the joint model, the joint classification and the method of global analysis are given in EC3 Part 1-8[2]. This is given in Table 10.1.

Table 10.1   Type of joint model

Method of global analysis	Classification of joint		
Elastic	Nominally pinned	Rigid	Semi-rigid
Rigid-plastic	Nominally pinned	Full-strength	Partial strength
Elastic-plastic	Nominally pinned	Rigid and full-strength	– Semi-rigid and partial-strength – Semi-rigid and full-strength – Rigid and partial-strength
Type of joint model	Simple	Continuous	Semi-continuous

According to the UK National Annex to EC3 Part 1-8[2], connections designed according to the principles given in the publication *Joints in steel construction – Simple connections*[29] may be classified as nominally pinned joints.

According to the UK National Annex to EC3 Part 1-8[2], connections designed according to the principles in the publication *Joints in steel construction – Moment connections*[30] may be classified according to the guidance in Section 2.5 of that document.

10.2 Connection design

This *Manual* covers connections in pin jointed and continuous frames. Connections for frames where the plastic hinge can form in the joint and semi-continuous frames are not covered.

The connection should be designed on a realistic, and consistent, assumption of the distribution of internal forces in the connection, which are in equilibrium with the externally applied loads. Each element in the connection must have sufficient resistance and deformation capacity.

The following points should be noted:
- The centroidal axes of the connected members should meet at a point; otherwise the effect of eccentricity of the connection should be taken into account in the design of the member.
- Bolts and welds in splice connections should be designed to carry all forces, except where provision is made for direct bearing where compressive forces are to be transferred by direct contact. It should be noted that rolled I and H sections can be prepared for direct bearing using a good quality saw (machining or milling is not required) and that full contact over the complete area is not required[64].
- EC3 Part 1-8[2] together with the UK National Annex provides for the use of ordinary bolts or high strength bolts in classes 4.6, 5.6, 8.8 and 10.9. Generally class 8.8 M20 bolts should be used for connections in members which will accommodate this size, and M16 used for smaller members. Heavily loaded connections may require M24 or M30 bolts, or the use of class 10.9 bolts. As far as possible only one size and class should be used on a project, see also Section 7.3.
- The local ability of the connected members to transfer the applied forces should be checked and stiffeners provided where necessary. Where the connection is not at the end of a member the design may have to take account of the combination of local and primary stresses in the member occurring at the connection.
- Bolted Shear Connections can be one of the following:
  Category A: Bearing type with appropriate shear and bearing resistance.
  Category B: Pre-loaded bolts for slip-resistance of the connection at serviceability limit state. They are used where slip would affect serviceability e.g. deflection of the structure. This may be the case for connections in trusses and bracing but in some cases the additional deflection is not significant and normal bolts can be used. They can be used for steelwork in building structures subjected to vibration or impact and are recommended for connections subject to load reversal.
  Category C: Pre-loaded bolts for slip-resistance of the connection at ultimate limit state i.e. where slip would affect the integrity of the structure.

- Bolted Tension Connections can be one of the following:
  Category D: Ordinary bolts, without pre-loading, may be used to resist wind forces and where there is little variation in other forces.
  Category E: Pre-loaded high strength bolts should be used where there is vibration or variable loading which would require fatigue consideration.
- The corrosion protection for the bolts should be compatible with the system used for the main frame.
- Where the steelwork is painted, the bolt assembly should include a washer under the part rotated during tightening to avoid damage to the painted surface.
- Where dissimilar metals are likely to be in contact in a moist environment, suitable isolators such as neoprene washers and sleeves should be incorporated to prevent bimetallic corrosion.

10.3 Bolts

10.3.1 Spacing and edge distances

End distances and spacing are illustrated in Figure 10.1 and Table 10.2 gives the minimum and maximum values according to EC3 Part 1-8[2]. The hole diameter is 2mm greater than the bolt diameter for M16 to M24 bolts and 3mm greater for bolts that are M27 and over. Where the minimum distances are used they will dictate the bearing resistance of the bolts (see Section 10.3.2).

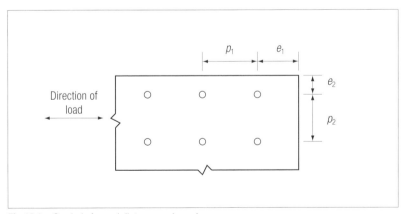

Fig 10.1 Symbols for end distance and spacing

10.3 Connections

Table 10.2 Bolt spacing and end distance

Requirement	Distance
Edge and end distances	
Minimum end e_1 and edge distance e_2 (with reduced bearing resistance)	$1.2d_0$ [a]
Maximum end and edge distances	
When exposed to weather or corrosion	$40\text{mm} + 4t$ [b]
When not exposed to weather or corrosion	No limit
Spacing	
Minimum spacing – direction of load p_1	$2.2d_0$
Minimum spacing – rows of fasteners p_2	$2.4d_0$
Maximum spacing	$14t$ or 200mm

Notes
a d_0 is the hole diameter.
b t is the minimum thickness of the connected parts.

10.3.2 Dimensions of holes

The dimensions of holes for bolted connections are given in BS EN 1090-2[64]. These are given in terms of nominal clearances. The nominal clearance is the difference between the nominal hole size and the nominal hole diameter. Values for nominal clearance for normal round holes, oversized holes and slotted holes are given in Table 10.3.

Table 10.3 Nominal clearances for bolts (mm)

Nominal bolt diameter (mm)	16	18	20	22	24	27 and over
Normal round holes		2			2	3
Oversize round holes		4			6	8
Short slotted holes (on the length)		6			8	10
Long slotted holes (on the length)	1.5 nominal bolt diameter					

10.3.3 Design resistance of ordinary bolts

Table 10.4 gives expressions for the resistance of ordinary bolts. The design material strengths given in Table 10.5 should be used.

The expression for tension resistance in Table 10.4 and the values of tension resistance in Table 10.6 are to be compared against a bolt tension that includes any prying action.

Table 10.4 Resistance of bolts

Design resistance for bolts	Formula
Shear resistance per shear plane for classes 4.6 and 8.8 (for threads in the shear plane)	$F_{v,Rd} = \dfrac{0.6 f_{ub} A_s}{\gamma_{M2}}$
Shear resistance per shear plane for class 10.9	$F_{v,Rd} = \dfrac{0.5 f_{ub} A_s}{\gamma_{M2}}$
N.B. when a shear connection is longer than $15d$ there is a reduction factor β_{LF} applied to the shear resistance	$\beta_{LF} = 1 - \dfrac{L_1 - 15d}{200\,d}$
Bearing resistance for normal round holes[a,b]	$F_{b,Rd} = \dfrac{k_1 \alpha_b f_u d t}{\gamma_{M2}}$
Tension resistance	$F_{t,Rd} = \dfrac{k_2 f_{ub} A_s}{\gamma_{M2}}$
Punching shear resistance	$B_{p,Rd} = 0.6 \pi d_m t f_u / \gamma_{M2}$
Combined shear and tension resistance	$\dfrac{F_{v,Ed}}{F_{v,Rd}} + \dfrac{F_{t,Ed}}{1.4 F_{t,Rd}} \leqslant 1.0$
Reduction factor β_p when using bolts through packing of thickness greater than $d/3$	$\beta_p = \dfrac{9d}{8d + 3t_p} \leqslant 1.0$

where:
f_{ub} is the ultimate tensile strength of the bolt
f_u is the ultimate tensile strength of the element
$F_{v,Ed}$ is the design shear force on the bolt
$F_{t,Ed}$ is the design tensile force on the bolt
A_s is the tensile stress area of the bolt
d is the diameter of the bolt
t is the minimum thickness of the connected parts
t_p is the thickness of packing

for edge bolts, k_1 is the smallest of $\dfrac{2.8 e_2}{d_0} - 1.7$, $\dfrac{1.4 p_2}{d_0} - 1.7$ and 2.5

for inner bolts, k_1 is the smaller of $\dfrac{1.4 p_2}{d_0} - 1.7$ and 2.5

α_b is the minimum of α_d, f_{ub}/f_u and 1.0

      for edge bolts, $\alpha_d = \dfrac{e_1}{3 d_0}$; for inner bolts $\alpha_d = \dfrac{p_1}{3 d_0} - \dfrac{1}{4}$

d_0 is the hole diameter
d_m is the mean of the across flats and across corners dimension of the nut or bolt (whichever is the smaller)
k_2 is 0.63 for countersunk bolts, otherwise it is 0.9
γ_{M2} is the partial factor, generally 1.25[c].

Table 10.4 Continued

Notes
a For oversized holes the bearing resistance is 0.8 times that for normal round holes and for slotted holes the bearing resistance perpendicular to the direction of the slot is 0.6 times that for normal round holes.
b The product of the coefficients k_1 and α_b can vary between 0.664 and 2.5. For single lap joints with one bolt row the product $k_1\alpha_b$ cannot be taken as more than 1.5.
c The UK National Annex to Part 1-8[2] gives a value of 1.25 for γ_{M2} but states that in certain circumstances to control deformation at serviceability a value of 1.5 would be more appropriate. This occurs when $\alpha_b=1.0$.

Table 10.6 gives values of tensile and shear resistance and the minimum ply thickness to avoid punching shear governing the tensile resistance for typical class 8.8 bolts.

Tables 10.7 and 10.8 give the bearing resistance for these bolts for typical edge and end distances. The factor α_b from Table 10.4 is less than 1 for these edge and end distances and the values in Tables 10.7 and 10.8 use the partial safety factor of 1.25 given in Table 10.4.

Table 10.5 Nominal yield strength and ultimate strengths for bolts (used as characteristic values in calculations)

Bolt class	4.6	8.8	10.9
Yield f_{yb} (N/mm^2)	240	640	900
Ultimate f_{ub} (N/mm^2)	400	800	1000

Table 10.6 Shear and tensile resistance for class 8.8 bolts

Bolt size	Tensile stress area (mm^2)	Tensile resistance (kN) at 576N/mm^2 (0.9f_{yb})	Shear resistance (kN) at 384N/mm^2 (0.6f_{yb}). Threads in shear plane		Minimum ply thickness to avoid punching shear (mm)	
			Single	Double	S275	S355
M16	157	90.4	60.3	121	5.8	5.1
M20	245	141.0	94.0	188	7.3	6.3
M24	353	203.0	136.0	271	8.7	7.6
M30	561	323.0	215.0	431	10.8	9.4

Note The tensile resistance given here does not apply for countersunk headed bolts.

Connections 10.3

Table 10.7 Bearing resistance of class 8.8 ordinary bolts: grade S275 material

Bolt size	Edge distance e_2 (mm)	End distance e_1 (mm)	Pitch p_1 (mm)	Gauge p_2 (mm)	Bearing resistance (kN) for f_u = 410N/mm² Thickness (mm) – plate passed through:							
					6	8	10	12	15	20	25	30
M16	25	35	50	50	44.7	59.6	74.5	89.3	112	*149*	*186*	*223*
M20	30	40	60	60	50.5	67.4	84.2	**101**	**126**	**168**	*211*	*253*
M24	35	50	70	70	62.6	83.5	104	125	**157**	**209**	**261**	*313*
M30	45	60	85	90	75.8	101	126	152	189	**253**	**316**	**379**

Notes
a Values in bold exceed the single shear resistance of the bolt.
b Values in italics exceed the double shear resistance of the bolt.

Table 10.8 Bearing resistance of class 8.8 ordinary bolts: grade S355 material

Bolt size	Edge distance e_2 (mm)	End distance e_1 (mm)	Pitch p_1 (mm)	Gauge p_2 (mm)	Bearing resistance (kN) for f_u = 470N/mm² Thickness (mm) – plate passed through:							
					6	8	10	12	15	20	25	30
M16	25	35	50	50	51.2	**68.3**	**85.4**	**102**	*128*	*171*	*213*	*256*
M20	30	40	60	60	57.9	77.2	**96.5**	**116**	**145**	*193*	*241*	*290*
M24	35	50	70	70	71.8	95.8	119.7	**144**	**180**	**239**	*299*	*359*
M30	45	60	85	90	86.9	115.8	144.8	174	**217**	**290**	**362**	*434*

Notes
a Values in bold exceed the single shear resistance of the bolt.
b Values in italics exceed the double shear resistance of the bolt.

10.3.4 Bolts in slip-resistant connections (Category B and C connections)

The formulae for slip-resistant (i.e. friction grip) connections are given in Table 10.9 and resistances are given in Table 10.10. They are based on class 8.8 bolt assemblies to BS EN 14399-3[65].

Bolts in accordance with BS EN 14399-4[66] can have thinner nuts and are outside the scope of this *Manual*. Bolts in Category B connections must also comply with the shear and bearing values given above. Bolts in Category C connections must comply with the bearing values given above. This is particularly important for bolts in Category B connections where the slip resistance is only at service load.

10.3 Connections

Table 10.9 Resistance of pre-loaded bolts in friction connections

Design resistance	Formulae
Slip resistance at ultimate limit state	$F_{s,Rd} = \dfrac{k_s n \mu}{\gamma_{M3}}(F_{p,C} - 0.8 F_{t,Ed})$
Slip resistance at serviceability limit state	$F_{s,Rd,ser} = \dfrac{k_s n \mu}{\gamma_{M3,ser}}(F_{p,C} - 0.8 F_{t,Ed,ser})$

where:
$F_{p,C}$ is the design preloading force = $0.7 f_{ub} A_s$
$F_{t,Ed}$ or $F_{t,Ed,ser}$ is the value of any applied tensile force on the connection
f_{ub} is the strength of the bolt, 800N/mm² for 8.8 bolts
A_s is the tensile stress area of bolt
k_s is a factor = 1.0 for nominal clearance holes
　　　　0.85 for oversize holes (24mm for M20, 30mm for M24)
　　　　0.63 for long slot holes (50mm for M20, 60mm for M24)
n is the number of friction interfaces
μ is the slip factor = 0.5 for blast cleaned surfaces with loose rust removed, not pitted
　　　　0.4 for blast cleaned and painted with an alkali zinc silicate coating of 50-80μm
　　　　0.3 for surfaces wire brushed or flame cleaned with loose rust removed
　　　　0.2 for surfaces as rolled
γ_{M3} is the partial safety factor with a value of 1.25 according to the UK National Annex
$\gamma_{M3,ser}$ is the partial safety factor with a value of 1.10 according to the UK National Annex.

Note The recommended preparation for surfaces is blast cleaning.

Table 10.10 Slip resistance of preloaded bolts

Resistances in kN for pre-loaded bolts in nominal clearance holes
Class 8.8 bolts to BS EN 14399-3, Class A surface (μ=0.5)

Bolt size	Tensile stress area (mm²)	Preload $F_{p,C}$ (kN)	Slip resistance			
			ULS		SLS	
			Single n=1	Double n=2	Single n=1	Double n=2
M16	157	87.9	35.2	70.3	40.0	79.9
M20	245	137.0	54.9	110.0	62.4	125.0
M24	353	198.0	79.1	158.0	89.9	180.0
M30	561	314.0	126.0	251.0	143.0	286.0

10.4 Welds

10.4.1 Fillet welds

It is normal practice in the UK to define fillet weld sizes by the leg length whereas in mainland Europe, reflected in EC3, the throat thickness is used. Designers and fabricators must ensure that they take cognisance of this difference and that the documentation for a project makes the assumption by the designer clear.

The effective length of a fillet weld is the length over which it is full size and according to EC3 Part 1-8[2] this may be taken as the total length less twice the effective throat thickness. A fillet weld with an effective length less than 30mm or less than six times its throat thickness should not be designed to carry load.

Unless access or the configuration of the joint makes this impractical, fillet welds finishing at the ends or sides of parts should not terminate at the corner but should be returned round the corner for twice the leg length (see Figure 10.2).

Fillet welds may be designed as continuous or intermittent, but intermittent welds should not be used in corrosive conditions or when subject to fatigue.

EC3 permits fillet welds to be used when the fusion faces form an angle between 60° and 120°.

A single fillet weld should not be subject to a bending moment or eccentric load about its longitudinal axis that would open the root of the weld.

The leg length should not be less than 4mm.

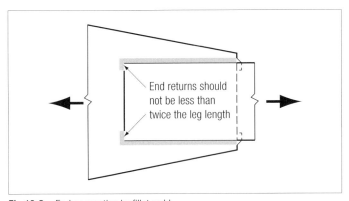

Fig 10.2 End connection by fillet weld

10.4 Connections

EC3 gives a Directional and Simplified method for weld design. According to the simplified method of design the resultant (vector sum) of all the connection forces on a weld, acting on a unit length of weld, must be less than the shear strength per unit weld. This can be calculated according to the formulae in Table 10.11 and typical resistances are given in Table 10.12. The resistances assume that the fusion faces form an angle of 90°; if it is greater than this the resistance will be reduced. For details of the Directional method see EC3 Part 1-8[2].

Table 10.11 Calculation of resistance of fillet welds

Design resistance	Formulae	
Shear resistance per unit length	$F_{w,Rd} = f_{vw,d} a$ where $f_{vw,d} = \dfrac{f_u/\sqrt{3}}{\beta_w \gamma_{M2}}$	(4.3 EC3-1-8) (4.4 EC3-1-8)

where:
f_u is the nominal ultimate tensile strength of the weaker part joined (see Table 2.6)
β_w is a correlation factor, 0.85 for S275 and 0.9 for S355 material
a is the throat thickness of the weld
γ_{M2} is the partial safety factor with a value of 1.25 according to the UK National Annex.

Note
In long joints there is a reduction factor on the design resistance of a fillet weld. This does not apply where the stress distribution along the weld corresponds with that in the parent metal. (e.g. the flange to web weld in a plate girder). In lap joints longer than $150a$ the reduction factor is: $\beta_{Lw,1} = 1.2 - 0.2 L_1/(150a)$ but $\beta_{Lw,1} \leqslant 1$
where L_1 is the overall length of the lap in the direction of the force transfer.

Table 10.12 Fillet weld resistances

Leg length (mm)	Throat thickness a (mm)	S275 Resistance at $f_{vw,d}$=222N/mm² (kN/mm)	S355 Resistance at $f_{vw,d}$=241N/mm² (kN/mm)
4	2.8	0.63	0.68
6	4.2	0.95	1.02
8	5.6	1.26	1.36
10	7.1	1.58	1.71
12	8.5	1.89	2.05
15	10.6	2.36	2.56

10.4.2 Butt welds

The weld may be a full penetration or a partial penetration butt weld.

Connections 10.4

A full penetration weld has a design strength equal to that of the weaker part joined.

A partial penetration butt weld has a design strength calculated for a fillet weld. As an alternative the strength can be calculated using the yield strength of the weaker part over a reduced depth equal to the throat thickness.

Situations where increased tension is produced at the root of single-sided partial penetration butt welds due to moments or eccentric loads should be avoided where possible. If such situations do occur the weld must be designed for the moment using the minimum depth of penetration.

Where a partial penetration Tee butt joint is reinforced with fillet welds as shown in Figure 10.3 the resistance should be determined as follows:
– Where the throat thickness a is greater than 70% of the minimum of s_1 and s_2 the resistance should be based on the minimum cross section area and the yield strength of the material
– Where the throat thickness is less than or equal to 70% of the minimum of s_1 and s_2 the resistance should be calculated using the procedure for a fillet weld with a throat thickness a.

For partial penetration welds, where the weld preparation is of the V or bevel type, the depth of preparation is usually more than the design throat thickness specified by the designer. Normally 3mm is added but smaller values can be justified by procedure trials. In that case it should not be taken as more than that which can be consistently achieved, ignoring weld reinforcement.

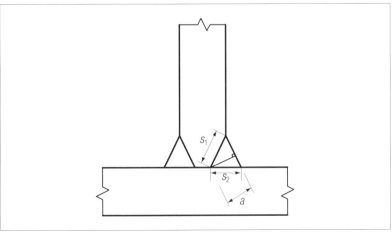

Fig 10.3 Tee butt weld with a pair of partial penetration butt welds with external fillets

11 Typical connection details

11.1 Introduction

This section describes typical connections for braced multi-storey buildings of simple construction, and for single-storey buildings, including portal frames.

A procedure is given for the design of each connection.

Most of the connections shown here have limited rotational stiffness. With such connections, it may be necessary to arrange construction so that floor restraints are in position in one area before proceeding further, or to provide temporary bracing, to ensure stability of the frame throughout its erection.

For more comprehensive methods of design, reference should be made to:
– *Joints in Steel Construction: Simple Connections* SCI/BCSA[29]
– *Joints in Steel Construction: Moment Connections* SCI/BCSA[30].

11.2 Column bases

11.2.1 General

Column bases should be of sufficient size, stiffness and strength to transmit safely the forces in the columns to the foundations. A uniform or linearly varying pressure distribution may be assumed in the calculation of the nominal bearing pressure between the baseplate and the supporting concrete.

11.2.2 Design of base plates

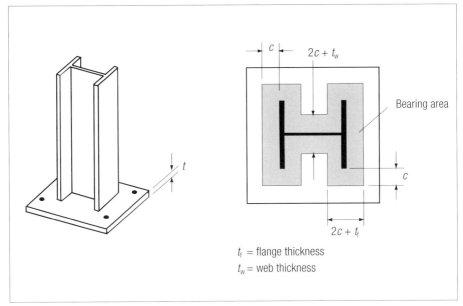

Fig 11.1 Simple base plate

Procedure
(1) Choose base plate size, length, breadth, thickness.
(2) Choose design strength f_{cd} – see Table 11.1.
(3) Calculate c from:

$$c = t \sqrt{\left(\frac{f_y}{3f_{jd}\gamma_{M0}}\right)}$$

where:
f_y is the yield strength of the base plate
f_{jd} is the bearing strength
γ_{M0} is the partial factor for steel (1.0).
(4) Check that c is within chosen base size, if not the bearing area is limited by the size of the plate.
(5) Calculate bearing area A_b as shown on Figure 11.1 and adjust for c as necessary.
NB: any overlap of the effective area between the flanges must be discounted, this occurs when $c > h - 2t_f$
(6) Calculate resistance of base = $A_b f_{jd}$
(7) Make provision for shear if adequate friction is not available based on coefficient of friction steel/concrete of 0.3.
(8) Design welds to accommodate any shear or tension.

11.2 Typical connection details

Table 11.1 Concrete design values and bearing strengths

	Strength Class				
f_{ck}/f_{cu}	C20/25	C25/30	C30/37	C35/45	C40/50
f_{cd} (N/mm²)	11.3	14.2	17.0	19.8	22.7

where:
f_{ck} is the concrete characteristic cylinder strength
f_{cu} is the concrete characteristic cube strength
f_{cd} is the design value of the concrete compressive strength given by the expression $\alpha_{cc} f_{ck}/\gamma_C$ which is the definition in EC2[35] rather than EC4[4]. According to the UK National Annex $\alpha_{cc} = 0.85$ and $\gamma_C = 1.5$
f_{jd} is the design bearing strength $= \beta_j F_{Rdu}/A_{c0}$
β_j is the joint coefficient, taken as 0.67 for when the characteristic strength of grout ≥ 0.2 of characteristic strength of concrete and grout thickness ≤ 0.2 of smallest width of base plate
F_{Rdu} is the concentrated design resistance force given in EC2[35] $\geq f_{cd} A_{c0}$
A_{c0} is the area loaded in compression.

Note
EC2 includes an expression to allow for the beneficial effects of confinement when the compression is applied to a small area and this can give significant benefits.

11.2.3 Design of bases (moment + axial load)

The design of bases under a combination of axial force and moment is relatively complicated and is not covered in detail in this *Manual*. There is particular guidance in EC3 Part 1-8 and also in the Access Steel document SN043[67]. The design relies on potential compression and/or tension forces either side of the column centreline. The compression is assumed to be centred on the column flange and the bearing resistance can be calculated using similar methods to Section 11.2.2. The tension is based on the bolts and checks are required on bolt embedment, bolt resistance, baseplate bending, welds to adjacent webs and flanges and the resistance of the adjacent parts of the column.

11.3 Beam-to-column and beam-to-beam connections in simple construction

Procedures are given below for three recognised types of connection used in simple construction. These are:
- web cleats
- flexible end plates
- fin plates.

Fin plates should not be used to connect Universal Beam sections which are greater than 610mm deep, unless it can be shown that the connection allows sufficient rotational movement to justify the simple method of design. This can normally be assumed if the span/depth ratio of the beam does not exceed 20 and the vertical distance between extreme bolts does not exceed 530mm.

In many cases horizontal ties will be required to provide robustness (see Section 3.7). Where steel beams are used to provide all or part of these ties the connections will need to be designed for the tie forces. It is recommended that the connections are designed for the applied shear and the tie force resistance is then calculated. For fin plate connections this can be taken as being equal to the applied shear. For the other connections it will usually be governed by the bending in the angle cleats or end plate. Increasing the tension resistance of web cleat and flexible end plate connections will often risk compromising their performance as simple connections and require the use of additional bolts over those required to carry the shear. This can often be avoided by using alternative paths for the tie forces.

11.3.1 Beam-to-column web cleats

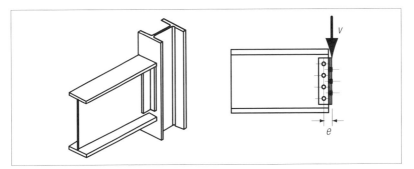

Fig 11.2 Beam to column web cleats

Procedure
(1) Choose size of pair of cleats (typically 90×90×10 angles) with sufficient length to take the required number of bolts. To provide torsional restraint to the end of the beam the angles should be at least 60% of the depth of the beam in length and positioned close to the top flange.

11.3 Typical connection details

(2) Bolts in the supporting member should be at reasonable cross centres (100mm plus beam web thickness).
(3) Calculate the size/number of bolts required in beam web to resist both shear V and moment V_e. See Section 10 for resistances.
(4) Calculate the size/number of bolts required between the cleats and the column. In this check the shear resistance of the bolts should be limited to 80% of that given in Section 10.
(5) Check shear and bearing value of cleats (both legs). The shear resistance should be based on the net area due to holes or 78% of the gross area.
(6) Check shear and bearing of beam web.
(7) Check for block shear failure as given in Section 11.3.6. It will be found in many cases, as determined from experience, that with the bolt arrangements adopted in the UK this mode of failure will rarely govern the design.
(8) Calculate the possible tie force based on the bending of the cleats and tension in the bolts to the supporting column.

11.3.2 Beam-to-beam web cleats

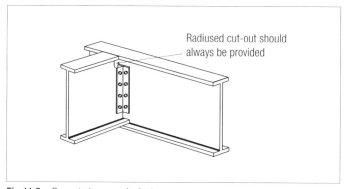

Fig 11.3 Beam to beam web cleats

Procedure
All as beam-to-column connection above plus:
(9) Check local shear and bearing of supporting beam web.
(10) Check reduced section of any notched beam for shear and bending, and the local stability of the notch.
(11) If the beam is unrestrained check the overall stability of any notched beam.

11.3.3 Beam-to-column flexible end plates

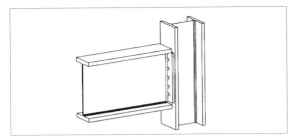

Fig 11.4   Beam to column flexible end plate

Procedure
(1) Choose plate 8 or 10mm thick using the same guidelines as given for the selection of angles in Section 11.3.1.
(2) Calculate the size/number of bolts required in plate to resist shear V. In this check the shear resistance of the bolts should be limited to 80% of that given in Section 10.
(3) Bolt cross centres should be reasonable (90mm to 140mm).
(4) See Section 10 for resistances.
(5) Check bearing on the endplate and supporting column.
(6) Check shear and bearing value of the end plate. The shear resistance should be based on the net area due to holes or 78% of the gross area.
(7) Check block shear in the end plate (see Section 11.3.6).
(8) Check shear resistance of beam web at the end plate.
(9) Choose fillet weld throat size to suit double length of weld.
(10) Calculate the possible tie force based on the bending of the end plate and tension in the bolts to the supporting column.
(11) Check column local resistance to resist the bolt tensions due to tie forces.

11.3.4 Beam-to-beam flexible end plates

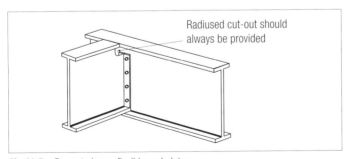

Fig 11.5   Beam to beam flexible end plate

Procedure
All as beam-to-column connection above plus checks similar to those listed in Section 11.3.2.

11.3 Typical connection details

11.3.5 Fin plates

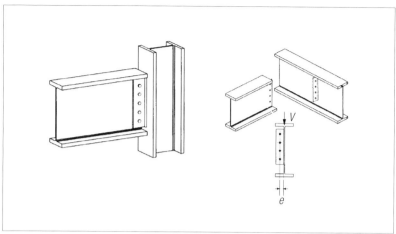

Fig 11.6 Fin plates

Procedure
(1) Choose fin plate size so that thickness is $\leq 0.5d$ (for S275) or $\leq 0.42d$ (for S355) where d is the bolt diameter and length ≥ 0.6 web depth.
(2) Calculate the size/number of bolts required in fin plate to resist both shear V and moment V_e.
(3) Normally use bolts of class 8.8 (see Section 10 for resistances).
(4) Check shear and bearing value of plate and bolts.
(5) Check shear resistance of beam web and end plate taking account of plain shear and block shear (see Section 11.3.6).
(6) Check fin plate for moment V_e.
(7) If beam is notched – check local stability of notch.
(8) Make the fillet weld throat size each side of the plate 0.5 (for S275) or 0.6 (for S355) times the plate thickness.
(9) Check local shear resistance of supporting beam or column web for beams supported on one or both sides.
Note See comment in Section 11.3 about fin plate connections for deep beams.

11.3.6 Block tearing

The possibility of failure in a connection caused by a block of material within the bolted area breaking away from the remainder of the section should be checked in certain cases. Some typical cases are shown in Figure 11.7.

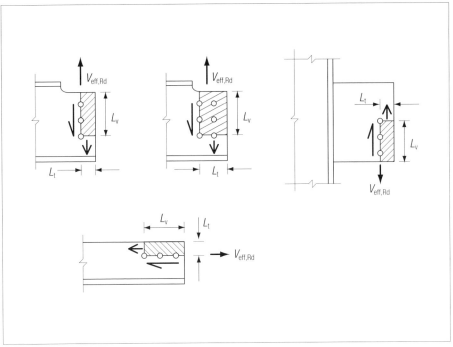

Fig 11.7 Typical cases of block tearing

The design block tearing resistance is given by the expression:

$V_{eff,Rd} = k_{bl} f_u A_{nt} / \gamma_{M2} + (1/\sqrt{3}) f_y A_{nv} / \gamma_{M0}$

where:
A_{nt} is the net area subject to tension = $(L_t - (n - 0.5)d_0)t_w$
A_{nv} is the net area subject to shear = $(L_v - (n - 0.5)d_0)t_w$
n is the number of bolt holes in the area being considered
d_0 is the diameter of the bolt hole
k_{bl} is 1.0 if the bolt group is symmetric and subject to concentric loading, otherwise it is 0.5

According to the UK National Annex γ_{M2} has a value of 1.25 and γ_{M0} a value of 1.0.

11.4 Column-to-column splices

Column splices should be located adjacent to and above floor levels and designed to meet the following requirements:
- They should be designed to hold the connected members in place.
- Wherever practicable the members should be arranged so that the centroidal axis of any splice material coincides with the centroidal axis of the member. If eccentricity is present, then the resulting forces must be taken into account.
- They should provide continuity of stiffness about both axes, and should resist any tension.
- They should provide the resistance to tensile forces to comply with the accidental action requirements of Section 3.7.
- Where the splice is not close to a position of restraint it should be designed for additional moments equal to:

$$M_{add} = \left(\left(\frac{1-\chi}{\chi}\right)\frac{N_{Ed}}{A}W_{pl} + \frac{\chi(\bar{\lambda}^2)C_m M_E}{(1-\chi(\bar{\lambda}^2))}\right)\sin\left(\frac{\pi x}{L}\right) \quad \text{about each axis}$$

together with

$$M_{add,z} = \left(\frac{1-\chi_{LT}}{\chi_{LT}}\right)\frac{W_{pl,z}}{W_{pl,y}}C_{my}M_{E,y}\sin\left(\frac{\pi x}{L}\right) \quad \text{about the minor axis}$$

where:
χ is the reduction factor for flexural buckling
χ_{LT} is the reduction factor for lateral torsional buckling
N_{Ed} is the axial compression in the column
A is the cross-section area of the column
W_{pl} is the plastic section modulus for the column about the relevant axis
C_m is the uniform moment factor C_{my} or C_{mz} from Table 5.3
M_E is the maximum applied moment
$M_{E,y}$ is the maximum applied moment about the major axis
$\bar{\lambda}$ is the relative slenderness
x is the distance of the splice from the floor connection
L is the storey height (system and effective length) of the column

The moment M_{add} should be calculated for each axis but it is only necessary to consider one axis at a time. Similarly, the contribution of $M_{E,y}$ to M_{add} and $M_{add,z}$ does not need to be considered simultaneously.

Typical connection details 11.4

11.4.1 Column bearing splices

Procedure
(1) Flange cover plates should project beyond the ends of each column by a length of not less than the upper column flange width or 225mm whichever is the greater.
(2) The thickness of flange cover plates should be the greater of half the thickness of the upper column flange or 10mm.
(3) When the upper and lower lengths are the same column serial size, nominal web cover plates may be used. They should incorporate at least four M20 class 8.8 bolts and have a width at least half the depth of the upper column.
(4) When the upper and lower lengths are of different sizes, then web cleats and a division plate should be used to give a load dispersal of 45°.
(5) Use M24 grade 8.8 bolts for splices in columns 305 series and above.
(6) The need for contact bearing should be noted on the drawings.

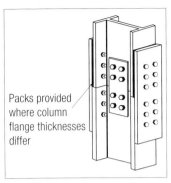

Packs provided where column flange thicknesses differ

Fig 11.8 Column bearing splice

EC3 Part 1-8[2] requires that the splice material can transmit 25% of the maximum force in the column and the above detailing rules will normally satisfy this rule.

It should be noted that the value of moment and/or sign of the axial force could cause a net tension in the cover plates. Tension may also exist to deal with the accidental force requirements of Section 3.7. In either case the tensile resistance of the cover plate based on gross and net area must be checked and the bolt group checked for shear and bearing. When checking the bolts, the reductions for long joints or packing given in Table 10.3 may need to be applied. Except when considering tension due to accidental forces, if the tensile force in the cover plate is greater than 10% of the resistance of the upper column flange then preloaded bolts designed to be non-slip at ULS should be used.

11.4.2 Column non-bearing splices

All the forces should be transmitted through the bolts and cover plates. A gap may be detailed between the sections to make it visually clear that a particular splice is non-bearing. Axial loads are shared between the web and flange cover plates in proportion to their areas and bending moments are deemed to be carried by the flanges. EC3 Part 1-8[2] requires that the moment about both axes should be not less than 25% of the moment resistance of the weaker section. It also requires that the splice can resist a shear equal to 2.5% of the shear resistance of the weaker section.

11.4 Typical connection details

Preloaded bolts should be provided if slip of the connection, which could lead to increased or excessive deflection, is not acceptable. This should be assumed to be the case if the splice is not adjacent to a restraint and has to sustain additional moments (see Section 11.4). In that case the bolts should be non-slip at ULS.

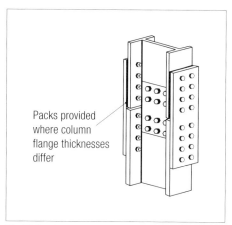

Packs provided where column flange thicknesses differ

Fig 11.9   Column non-bearing splice

Procedure
(1) Choose the splice plate sizes, number and type of bolts. A useful starting point is to follow the recommendations for bearing splices.
(2) Calculate the forces in the flange and web plates due to axial force (including any tying), moments and shears.
(3) Check the resistance of the plates using the net area for shear, the gross or net area for any tension and the gross area for compression.
(4) Calculate forces in bolts arising from axial loads, moments and shears.
(5) Check bolt strengths in single shear, double shear or bearing depending on the arrangement of cover plates. Take account of any reductions for long joints or pack as necessary.
(6) Check the bearing stress in the flanges and splice covers.

11.5 Portal frame eaves connections

11.5.1 General

The analysis and design of a portal frame eaves connection is complicated and is best carried out using software or tables of standard solutions. The Access Steel document SN041[68] includes a simplified method of calculation. The method given in this section includes certain assumptions and will give a resistance that may be very conservative.

Where portal frames are analysed by the plastic method the following should be adopted:
- A plastic hinge should not be allowed to form in the joint.
- A long haunch should be treated as a frame member.
- The depth of haunch chosen may conveniently be arranged so that two haunches may be cut from a single length of the same section as shown in Figure 11.10. This is assumed to be the same section as that of the rafter.
- The angle of the haunch flange to the flange of the rafter α should not be greater than 45°.
- The design method shown here is for bolt rows each being composed of two bolts only.
- Only the bolts in the top half of the connection should be considered as tension fasteners.
- It is assumed that the column is a UB or UC section.
- It is assumed that the thickness of the end plate on the beam is at least equal to the column flange.
- It is assumed that there is no stiffener to the column flange in the tension zone.
- It is assumed that the end plate is flush and that it projects below the bottom of the haunch by a distance not less than the end plate thickness plus the leg length of the fillet weld.
- The distance from the top row of bolts to the end of the column is at least half the pitch of the bolt rows.

The Eurocode uses three similar symbols related to these connections and these can be confusing. They are:
- $F_{t,Rd}$ - tension resistance of an individual bolt
- $F_{T,Rd}$ - tension resistance of an equivalent T-stub
- $F_{tr,Rd}$ - tension resistance of a bolt row.

11.5 Typical connection details

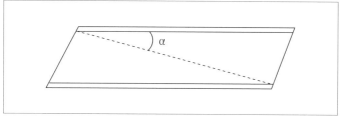

Fig 11.10 Cutting two sections of haunch from a length of rafter

11.5.2 Geometry – eaves haunch

The geometry of the eaves connection is shown in Figure 11.11 and the expression for calculating the dimensions m and e are shown in Table 11.2.

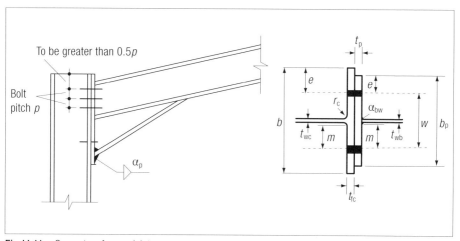

Fig 11.11 Geometry of eaves joint

Table 11.2 Expressions for m and e

For column flange	For end plate
$m = \dfrac{w}{2} - \dfrac{t_{wc}}{2} - 0.8r_c$	$m = \dfrac{w}{2} - \dfrac{t_{wb}}{2} - 0.8a_{bw}\sqrt{2}$
$e = \dfrac{b}{2} - \dfrac{w}{2}$	$e = \dfrac{b_p}{2} - \dfrac{w}{2}$

11.5.3 Resistance of column flange in bending

The three modes of failure shown in Figure 11.12 have to be considered. Expressions for the resistance of the modes are given in Table 11.3.

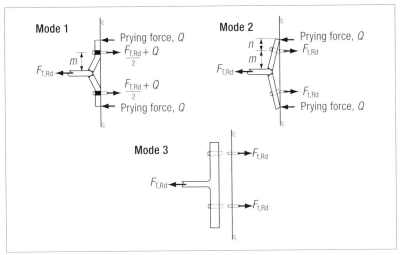

Fig 11.12   Modes of failure of T-stubs

Table 11.3   Resistance of T-stub

Mode	Failure type	Resistance
Mode 1	Column flange yielding (without backing plates)	$F_{T,Rd} = \dfrac{4M_p}{m}$
Mode 2	Bolt failure with flange yielding	$F_{T,Rd} = \dfrac{2M_p + n(\sum F_{t,Rd})}{m + n}$
Mode 3	Bolt failure	$F_{T,Rd} = \sum F_{t,Rd}$
where: $\sum F_{t,Rd}$ = sum of design tension resistances of the bolts in a row i.e. $2F_{t,Rd}$ n = minimum of e for column and end plate but not more than $1.25m$ M_p = plastic moment resistance of an equivalent T-stub = $\dfrac{L_{eff} t_{fc}^2 f_y}{4\gamma_{M0}}$ L_{eff} = effective length in equivalent T-stub (see Section 11.5.4) t_{fc} = column flange thickness f_y = design strength of column flange.		
Note To reduce the number of calculations the resistance of the T-stub ($F_{T,Rd}$) should be calculated using the maximum value of m from the column or the beam.		

11.5 Typical connection details

11.5.4 Effective length of equivalent T-stub

In calculating the effective length of the equivalent T-stub, multiple rows of bolts should really be considered and where necessary, the effect of flanges and stiffeners on the effective length. To simplify the calculation, the resistance of each bolt row is considered individually and based on the assumptions at the beginning of this section that the effective length can be taken as the pitch of the bolt rows p.

11.5.5 Web tension in beam or column

Check that the potential load from the bolt row does not exceed the resistance of the web or weld:

for the column web: $F_{t,wc,Rd} = \dfrac{L_{eff} t_{wc} f_{y,wc}}{\gamma_{M0} \sqrt{(1 + 1.3(L_{eff}/h)^2)}}$

for the beam web: $F_{t,wb,Rd} = L_{eff} t_{wb} f_{y,wb}/\gamma_{M0}$

where:
L_{eff} is the effective length of the web taken as equal to bolt pitch
t_w is the thickness of the web
f_y is the design strength
h is the depth of the column section

11.5.6 Final resistance moment

The resistance moment is calculated by adding the moment due to the tension in each bolt row about the centre of compression. The centre of compression may be taken as the centre of the bottom flange of the beam or haunch.

Consider the minimum bolt row forces (flange bending, bolt failure or web tension) on the column side using the failure mechanisms given in the earlier text. This gives the tensile resistance $F_{tr,Rd}$ of the bolt row, it is assumed that the bolt pitch remains constant and so this resistance is applicable to all rows. This has been calculated for bending of the column flange, but the resistance based on bending of the end plate will be at least this value, assuming the end plate is at least as thick as the column flange.

According to the UK National Annex to EC3 Part 1-8[2], provided either the bolt row force $F_{tr,Rd}$ does not exceed 1.9 times the tension resistance of an individual bolt $F_{t,Rd}$, or the column flange thickness does not exceed the value given in Table 11.4 a plastic distribution of bolt forces can be assumed as shown in Figure 11.13.

Typical connection details 11.5

Table 11.4 Maximum column flange thickness in mm for plastic distribution of bolt forces

Bolt size	Grade of column flange	
	S275	S355
M20	18.3	16.0
M24	21.9	19.2
M30	27.5	24.0

If both of these conditions do not apply a triangular distribution of forces (as shown in Figure 11.13) should be assumed. The bolt force should be proportional to the distance from the centre of compression.

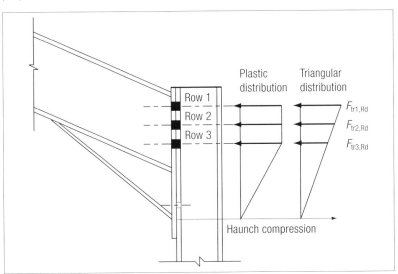

Fig 11.13 Distribution of bolt forces

11.5.7 Compression checks

The compression force acting at the bottom of the haunch, may be taken as the sum of the bolt row forces calculated as above ($\Sigma F_{tr,Rd}$). The resulting reactions on the column must be less than the design resistance of the column. The design resistance of an unstiffened web may be calculated using the following expression:

$$F_{c,wc,Rd} = \frac{0.7 \rho b_{eff,c,wc} t_{wc} f_{y,wc}}{\gamma_{M1} \sqrt{(1 + 1.3(b_{eff,c,wc}/h)^2)}}$$

11.5 Typical connection details

The 0.7 factor assumes that the longitudinal stress in the column web due to moment and axial force is at yield. If the stress is below $0.7f_{y,wc}$ where $f_{y,wc}$ is the yield strength of the column web, this can be increased to 1.0.

The effective width of the column web is given by the expression:

$$b_{eff,c,wc} = t_{fb} + 2a_p\sqrt{2} + 5(t_{fc} + r_c) + 2t_p$$

The last term in this expression is based on a 45° spread through the end plate and assumes that the end plate projects below the end of the weld to the bottom flange of the haunch by at least t_p. For dimensions see Figure 11.11, a_p is the throat size of the weld between the flange of the haunch and the end plate.

The reduction factor for plate buckling is given by the expression:

- if $\bar{\lambda}_p \leq 0.72$: $\quad \rho = 1.0$
- if $\bar{\lambda}_p > 0.72$: $\quad \rho = (\bar{\lambda}_p - 0.2)/\bar{\lambda}_p^2$

The plate slenderness is given by the expression:

$$\lambda_p = 0.932\sqrt{\frac{b_{eff,c,wc}d_{wc}f_{y,wc}}{Et_{wc}^2}}$$

The dimension d_{wc} is the clear depth of the web i.e. $(h - 2(t_f + r_c))$, other dimensions are as shown in Figure 11.11.

There is also compression on the rafter at the end of the haunch and the same procedure can be used to check the component of force perpendicular to the rafter.

11.5.8 Column web panel shear

The column panel must be capable of resisting the shear forces arising from the tensile forces in the bolts. This is V_{Ed} and should be less than 90% of the plastic resistance of the web to shear $V_{pl,Rd}$. In the event of the shear being greater than this value stiffening should be provided either by the use of diagonal stiffeners, or the use of a reinforcing web plate welded to the sides of the web. The SCI/BCSA publication on moment connections[30] includes diagrams of options. The plastic resistance of the web may be derived as Section 4.1.2(4).

11.5.9 Vertical shear

The welds connecting the haunch to the end plate should be checked for the shear forces in addition to the tensile forces due to the moments.

The bolts connecting the end plate to the face of the column should be checked for shear as well as tension. In most practical portal frames it will be found that the bolts needed to hold the bottom of the haunch in place will be capable of carrying the shear without the necessity of deriving the resistance of those at the top in combined tension and shear.

11.5.10 Weld sizes

The welds to the rafter flange and web in the tension zone should conservatively be full strength. To achieve this, the throat thickness of the fillet weld on both sides of the flange or web should at least be equal to the following:
for S275 material: $a \geqslant 0.5t$
for S355 material: $a \geqslant 0.6t$

If the compression flange of the beam or haunch has been properly sawn so that it can bear against the end plate a nominal weld should be sufficient and fillet welds with a throat thickness of 5mm each side of the flange are recommended for normal situations.

11.5.11 Summary of procedure

(1) Choose the number of bolts and the geometry for a trial connection making end plate thickness equal to or greater than the column flange thickness.
(2) Calculate the potential resistances of bolt rows in tension taking account of:
 – column flange bending
 – column web tension
 – and if necessary beam web tension for rows not adjacent to beam flange.
(3) Calculate the total resistance of the bolt rows in tension, which equates to the total compressive force in the connection after allowance is made for axial forces.
(4) Calculate the moment of resistance of the connection.
(5) Re-configure connection if the moment of resistance is less than that required and repeat steps (1) to (4). It may not be possible to obtain the required moment using the simplified procedure given here.
(6) Check column compression resistance, providing compression stiffeners if found necessary.
(7) Check column web panel shear resistance, providing compression stiffeners if found necessary.
(8) Check beam compression flange bearing.
(9) Check shear resistance of connection.
(10) Design the welds.

11.6 Resistance to transverse forces i.e web bearing and buckling

This check should be carried out when concentrated loads are applied to unstiffened webs and might cause bearing failure or buckling of the web. Although the expression given in Section 11.5.7 would appear relevant, the resistance should be calculated according to EC3 Part 1-5[28]. This gives an expression for the resistance to transverse loads that depends on the length of stiff bearing, whether the load is at the end or not and the position of transverse stiffeners.

The following expressions are conservative simplifications of the full guidance in Part 1-5. They apply to UB and UC sections with no transverse stiffeners and where the concentrated load is applied to one flange only with no stiff bearing length.

The design resistance to transverse forces should be taken as:

$$F_{Rd} = \frac{f_{yw} L_{eff} t_w}{\gamma_{M1}}$$ (6.1 EC3-1-5)

where:
f_{yw} is the yield strength of the web
t_w is the web thickness
γ_{M1} is the partial safety factor, given a value of 1.0 in the UK National Annex
L_{eff} is the effective length for resistance to transverse forces, this can conservatively be taken as:

$$t_f \sqrt{\frac{0.5 b_f}{t_w}}$$ for loads applied at the end of the beam

$$2t_f \left(1 + \sqrt{\frac{b_f}{t_w}}\right)$$ for loads applied away from the end of the beam

where:
b_f is the width of the flange
t_f is the flange thickness.

References

1 *BS EN 1993-1-1: 2005: Eurocode 3: Design of steel structures – Part 1-1: General rules and rules for buildings.* London: BSI, 2005.

2 *BS EN 1993-1-8: 2005: Eurocode 3: Design of steel structures – Part 1-8: Design of joints.* London: BSI, 2005.

3 *BS EN 1993-1-10: 2005: Eurocode 3: Design of steel structures – Part 1-10: Selection of steel for fracture toughness and through-thickness properties.* London: BSI, 2005.

4 *BS EN 1994-1-1: 2004: Eurocode 4: Design of composite steel and concrete structures – Part 1-1: General rules and rules for buildings.* London: BSI, 2004.

5 *BS EN 1998-1: 2004: Eurocode 8: Design of structures for earthquake resistance – Part 1: General rules, seismic actions and rules for buildings.* London: BSI, 2004.

6 Steel Construction Institute. *Steel building design: design data. SCI Publication P363.* Ascot: SCI, 2009.

7 *The Construction (Design and Management) Regulations.* Norwich: The Stationery Office, 2007 (SI 2007/320).

8 Health and Safety Commission. *Managing health and safety in construction: Construction (Design and Management) Regulations 2007. L144.* [Sudbury]: HSE Books, 2007.

9 Construction Skills. *Guidance* [for CDM Regulations]. Available at http://www.cskills.org/supportbusiness/healthsafety/cdmregs/guidance/index.aspx [Accessed: 5 August 2009].

10 British Constructional Steelwork Association. *Allocation of design responsibilities in constructional steelwork. BCSA publication 45/07.* London: BCSA, 2007.

11 Lemaire, V. *Scheme development: Movement joints in steel buildings. Access Steel document SS017a-EN-EU.* Available at http://www.access-steel.com. [Accessed: 5 August 2009].

12 *BS EN 1991-1-1: 2002: Eurocode 1: Actions on structures – Part 1-1: General actions – Densities, self weight, imposed loads for buildings.* London: BSI, 2002.

13 *BS EN 1991-1-3: 2003: Eurocode 1: Actions on structures – Part 1-3: General actions – Snow loads.* London: BSI, 2003.

14 *BS EN 1991-1-4: 2005: Eurocode 1: Actions on structures – Part 1-4: General actions – Wind loads.* London: BSI, 2005.

15 *BS EN 1997-1: 2004: Eurocode 7: Geotechnical design – General rules.* London: BSI, 2004.

16 *BS EN 1990: 2002: Eurocode: Basis of structural design.* London: BSI, 2002.

17 Institution of Structural Engineers *Manual for the design of building structures to Eurocode 1 and basis of structural design.* London: Institution of Structural Engineers, 2010.

18 Willford, M.R. and Young, P. *A Design guide for footfall induced vibration of structures. CCIP-106.* Camberley: The Concrete Centre, 2006.

References

19 Smith, A.L. et al. *Design of floors for vibration: a new approach. SCI Publication P354.* Ascot: SCI, 2007.

20 Department of Health. *Specialist services: acoustics. Health Technical Memorandum 08-01.* Norwich: The Stationery Office, 2008.

21 Institution of Structural Engineers et al. *Dynamic performance requirements for permanent grandstands subject to crowd action: recommendations for management design and assessment.* London: Institution of Structural Engineers, 2008.

22 *BS EN 10025: Hot rolled products of structural steel* [6 parts].

23 *BS EN 10210: Hot finished structural hollow sections of non-alloy and fine grain steel* [2 parts].

24 *BS EN 10219: Cold formed welded structural hollow sections of non-alloy and fine grain steel* [2 parts].

25 *BS EN 1993-1-2: 2005: Eurocode 3: Design of steel structures – Part 1-2: General rules – Structural fire design.* London: BSI, 2005.

26 *PD 6695-1-10: 2009: Recommendations for the design of structures to BS EN 1993-1-10.* London: BSI, 2009.

27 *BS EN 10164: 2004: Steel products with improved deformation properties perpendicular to the surface of the product – Technical delivery conditions.* London: BSI, 2004.

28 *BS EN 1993-1-5: 2006: Eurocode 3: Design of steel structures – Part 1-5: Plated structural elements.* London: BSI, 2006.

29 Steel Construction Institute and British Constructional Steelwork Association. *Joints in steel construction: simple connections. SCI Publication P212.* Ascot: SCI, 2002.

30 Steel Construction Institute and British Constructional Steelwork Association. *Joints in steel construction: moment connections. SCI Publication P207.* Ascot: SCI, 1995.

31 Couchman, G.H. *Design of semi-continuous braced frames. SCI Publication P183.* Ascot: SCI, 1997.

32 Salter, P.R. et al. *Wind-moment design of low rise frames. SCI Publication P263.* Ascot: SCI, 1999.

33 Hicks, S. et al. *Comparative structure cost of modern commercial buildings. SCI Publication P137.* 2nd ed. Ascot: SCI, 2004.

34 *BS EN 1995-1-1: 2004: Eurocode 5: Design of timber structures – Part 1-1: General. Common rules and rules for buildings.* London: BSI, 2004.

35 *BS EN 1992-1-1: 2004: Eurocode 2: Design of concrete structures – Part 1-1: General rules and rules for buildings.* London: BSI, 2004.

36 Department for Communities and Local Government. *The Building Regulations 2000. Approved Document B: Fire Safety. Vol 1: Dwellinghouses; Vol 2: Buildings other than dwellinghouses.* 2006 ed. London: NBS, 2007.

37 Ham, S.J. et al. *Structural fire safety: a handbook for architects and engineers. SCI Publication P197.* Ascot: SCI, 1999.

38 Morris, W.A. et al. *Guidelines for the construction of fire resisting structural elements. BRE Report BR128.* Garston: BRE, 1998.

39 Association for Specialist Fire Protection. *Fire protection for structural steel in buildings.* 4th ed. Aldershot: ASFP, 2007.

References

40 Institution of Structural Engineers. *Introduction to the fire safety engineering of structures*. London: Institution of Structural Engineers, 2003.

41 Institution of Structural Engineers. *Guide to the advanced fire safety engineering of structures*. London: Institution of Structural Engineers, 2007.

42 *BS EN ISO 12944: Paints and varnishes – Corrosion protection of steel structures by protective paint systems* [8 parts].

43 Gedge, G. and Whitehouse, N. *New paint systems for the protection of constructional steelwork. CIRIA Report 174*. London: CIRIA, 1997.

44 British Constructional Steelwork Association and Galvanizers Association. *Galvanizing structural steelwork. BCSA Publication 40/05*. London: BCSA, 2005.

45 *BS EN 1991-1-7: 2006: Eurocode 1 – Actions on structures – Part 1-7: General actions – Accidental actions*. London: BSI, 2006.

46 *BS 4-1: 2005: Structural steel sections. Specification for hot rolled sections*. London: BSI, 2005.

47 Lim J. *NCCI: Determination of non-dimensional slenderness of I and H sections. Access Steel document SN002a-EN-EU*. Available at: http://www.access-steel.com [Accessed 5 August 2009].

48 Way, J. *NCCI: Effective length and destabilizing load parameters for beams and cantilevers – common cases. Access Steel document SN009a-EN-EU*. Available at: http://www.access-steel.com [Accessed 9 August 2009].

49 *BS EN 1991-1-6: 2005: Eurocode 1 – Actions on structures – Part 1-6: General actions – Actions during execution*. London: BSI, 2005.

50 *BS 8500: Concrete. Complementary British Standard to BS EN 206-1* [2 parts].

51 *BS EN 206-1: 2000: Concrete. Specification, performance, production and conformity*. London: BSI, 2000.

52 *BS 4449: 2005: Steel for the reinforcement of concrete – Weldable reinforcing steel – Bar, coil and decoiled product. Specification*. London: BSI, 2005.

53 *BS 4483: 2005: Steel fabric for the reinforcement of concrete. Specification*. London: BSI. 2005.

54 Heywood, M.D. *NCCI: Determination of moments on columns in simple construction. Access Steel document SN005a-EN-EU*. Available at: http://www.access-steel.com [Accessed 5 August 2009].

55 Narboux L. *NCCI: Verification of columns in simple construction – a simplified criterion. Access Steel document SN048a-EN-GB*. Available at: http://www.access-steel.com [Accessed 5 August 2009].

56 Banfi, M. Simplified expressions for compression and bending. *The Structural Engineer*, 27(21), 4 November 2008, pp24-29.

57 BRE. *Industrial platform floors: mezzanine and raised storage. BRE Digest 437*. Garston: BRE, 1999.

58 British Constructional Steelwork Association. *Guide to Eurocode load combinations for steel structures* [in preparation].

59 Raven, G.K. and Heywood, M.D. *Single storey buildings. SCI Publication P347*. Ascot: SCI, 2006.

References

60 Simms, W.I. and Newman, G.M. *Single storey steel framed buildings in fire boundary conditions. SCI Publication P313*. Ascot: SCI, 2002.

61 Chica, J.A. *NCCI:Column base stiffness for global analysis. Access Steel document SN045a-EN-EU*. Available at: http://www.access-steel.com [Accessed 5 August 2009].

62 Oppe, M. *NCCI: Simple methods for second order effects in portal frames. Access Steel document SN033a-EN-EU*. Available at: http://www.access-steel.com [Accessed 5 August 2009].

63 Bureau, A. *NCCI: Practical deflection limits for single storey buildings. Access Steel document SN035a-EN-EU*. Available at: http://www.access-steel.com [Accessed 5 August 2009].

64 *BS EN 1090-2: 2008. Execution of steel structures and aluminium structures. Part 2. Technical requirements for steel structures.* London: BSI, 2008.

65 *BS EN 14399-3: 2005: High-strength structural bolting assemblies for preloading. Part 3: System HR. Hexagon bolt and nut assemblies.* London: BSI, 2005.

66 *BS EN 14399-4: 2005: High-strength structural bolting assemblies for preloading. Part 4: System HV. Hexagon bolt and nut assemblies.* London: BSI, 2005.

67 Ryan, I. *NCCI: Design of fixed column base joints. Access Steel document SN043a-EN-EU*. Available at: http://www.access-steel.com [Accessed 5 August 2009].

68 Grijaivo, J. *NCCI: Design of portal frame eaves connections. Access Steel document SN041a-EN-EU*. Available at: http://www.access-steel.com [Accessed 5 August 2009].

69 *BS EN ISO 13918: 2008: Welding – Studs and ceramic ferrules for arc stud welding*. London: BSI, 2008.

70 Couchman, G and Smith, A. *NCCI: Resistance of headed stud shear connectors in transverse sheeting (GB) Access Steel document PN001a-GB*. Available at: http://www.steelbiz.org [Accessed 15 March 2010].

71 Smith, A. *NCCI: Modified limitation on partial shear connection in beams for buildings (GB) Access Steel document PN002a-GB*. Available at: http://www.steelbiz.org [Accessed 15 March 2010].

72 *PD 6695-1-9: Recommendations for the design of structures to BS EN 1993-1-9*. London: BSI, 2008.

Appendix A Moment resistances of UB sections

Moment resistance $M_{c,Rd,y}$ for fully restrained beams, critical effective length for maximum $M_{c,Rd,y}$ and buckling resistance moment for UB sections are given in Tables A1 and A2 for S275 and in Tables A3 and A4 for S355 steel.

Table A1   S275 Steel – 457 deep and above

Serial size	Mass (kg/m)	I_y (cm^4)	$M_{c,Rd,y}$ (kNm)	Critical value of L_E (m)	Buckling resistance moment (kNm) for effective length of L_E (m)					
					4	5	6	7	8	9	10
1016×305	487 [a]	1021884	5918	2.85	5226	4717	4280	3907	3589	3315	3080
	437 [a]	910322	5296	2.79	4622	4148	3741	3394	3099	2847	2633
	393 [a]	807503	4727	2.73	4074	3635	3257	2935	2663	2433	2238
	349 [a]	723131	4397	2.66	3734	3309	2941	2630	2368	2148	1963
	314 [a]	644063	3935	2.62	3307	2915	2576	2288	2048	1847	1679
	272 [a]	553974	3399	2.60	2835	2486	2182	1924	1709	1530	1381
	249 [a]	481192	3008	2.52	2460	2140	1863	1629	1435	1276	1145
	222 [a]	407961	2599	2.43	2077	1791	1544	1338	1169	1031	919
914×419	388	719635	4681	3.88	4650	4398	4151	3907	3669	3439	3222
	343	625780	4102	3.81	4058	3830	3604	3380	3159	2946	2744
914×305	289	504187	3331	2.69	2986	2726	2471	2232	2015	1824	1659
	253	436305	2900	2.65	2580	2343	2110	1890	1692	1519	1371
	224	376414	2527	2.59	2226	2010	1796	1596	1416	1260	1129
	201	325254	2213	2.52	1927	1729	1533	1350	1187	1049	932
838×292	227	339704	2426	2.57	2137	1934	1737	1553	1389	1246	1125
	194	279175	2025	2.50	1758	1577	1401	1237	1093	970	867
	176	246021	1804	2.44	1550	1383	1219	1068	937	825	733
762×267	197	239957	1899	2.35	1614	1444	1283	1138	1013	908	819
	173	205282	1642	2.30	1378	1222	1075	943	831	737	660
	147	168502	1366	2.23	1126	989	859	744	647	568	504
	134	150692	1277	2.16	1033	898	771	661	571	497	438
686×254	170	170326	1492	2.27	1250	1114	987	875	778	698	631
	152	150355	1325	2.23	1099	973	854	750	662	589	529
	140	136267	1208	2.20	993	874	762	665	583	516	461
	125	117992	1058	2.15	857	748	646	558	484	425	377

Appendix A

Table A1   S275 Steel – 457 deep and above (Continued)

Serial size	Mass (kg/m)	I_y (cm⁴)	$M_{c,Rd,y}$ (kNm)	Critical value of L_E (m)	Buckling resistance moment (kNm) for effective length of L_E (m)					
					4	5	6	7	8	9	10
610×305	238	209471	1984	2.95	1843	1716	1594	1480	1374	1279	1192
	179	153024	1470	2.84	1345	1239	1135	1037	947	866	794
	149	125876	1217	2.79	1105	1011	920	832	752	680	617
610×229	140	111777	1098	2.06	880	774	678	597	529	474	429
	125	98610	974	2.02	771	672	583	508	447	398	357
	113	87318	869	1.99	679	587	504	436	380	335	299
	101	75780	792	1.91	601	511	433	369	318	278	247
610×178	100 a	72528	738	1.50	417	339	282	241	210	186	167
	92 a	64577	691	1.44	370	296	244	206	178	157	140
	82 a	55869	603	1.41	312	247	201	168	145	126	112
533×312	273 a	198971	2085	3.20	2016	1933	1851	1772	1694	1618	1545
	219 a	151281	1621	3.06	1551	1476	1402	1327	1253	1181	1113
	182 a	123474	1335	2.98	1268	1201	1132	1062	992	924	861
	150 a	100840	1099	2.92	1038	978	915	851	786	724	666
533×210	138 a	86088	957	0.67	754	667	590	525	472	427	390
	122	76043	847	0.60	658	575	504	445	396	356	323
	109	66822	750	0.53	572	496	429	375	331	295	267
	101	61519	692	1.86	524	451	388	336	295	263	236
	92	55227	649	1.80	477	405	344	295	257	226	202
	82	47539	566	1.76	408	342	287	243	210	184	163
533×165	85 a	48505	558	1.43	304	248	207	178	156	139	125
	74 a	41058	497	1.35	252	201	166	141	122	108	97
	66 a	35044	429	1.32	208	164	134	112	97	85	76
457×191	161 a	79779	1001	1.94	820	747	683	627	579	536	499
	133 a	63841	814	1.85	642	575	517	468	426	391	361
	106 a	48873	633	1.78	479	419	369	327	293	265	242
	98	45727	592	1.77	443	385	337	297	265	239	218
	89	41015	534	1.75	394	339	294	257	228	204	185
	82	37051	504	1.68	359	305	260	226	198	176	159
	74	33319	455	1.67	319	269	228	195	170	151	135
	67	29380	405	1.64	278	231	194	165	143	125	112

Appendix A

Table A1   S275 Steel – 457 deep and above (Continued)

Serial size	Mass (kg/m)	I_y (cm^4)	$M_{c,Rd,y}$ (kNm)	Critical value of L_E (m)	Buckling resistance moment (kNm) for effective length of L_E (m)						
					4	5	6	7	8	9	10
457×152	82	36589	480	1.39	307	258	220	192	170	152	138
	74	32674	431	1.37	267	222	187	162	142	127	115
	67	28927	400	1.32	235	191	160	137	119	106	95
	60	25500	354	1.30	201	162	133	113	98	86	76
	52	21369	301	1.26	163	128	104	87	75	65	56

Note
a These sections are part of the Corus Advance range but are not in BS4[46]

Table A2   S275 Steel – 406 deep and below

Serial size	Mass (kg/m)	I_y (cm^4)	$M_{c,Rd,y}$ (kNm)	Critical value of L_E (m)	Buckling resistance moment (kNm) for effective length of L_E (m)						
					2	2.5	3	4	5	6	7
406×178	85 [a]	31703	459	1.68	441	414	387	337	292	256	226
	74	27310	413	1.61	392	365	339	289	246	212	185
	67	24331	370	1.58	350	325	300	254	214	182	157
	60	21596	330	1.57	310	288	265	222	185	156	133
	54	18722	290	1.53	271	250	229	189	155	129	109
406×140	53 [a]	18283	284	1.23	245	221	197	156	126	105	89
	46	15685	244	1.21	210	187	166	129	102	84	71
	39	12508	199	1.16	167	148	129	97	75	61	51
356×171	67	19463	333	1.58	315	294	273	235	202	175	154
	57	16038	278	1.55	261	242	223	188	159	136	118
	51	14136	246	1.52	230	213	195	163	136	114	98
	45	12066	213	1.49	197	182	166	136	112	93	79
356×127	39	10172	181	1.08	148	130	113	87	69	57	49
	33	8249	149	1.04	119	103	89	66	52	42	35
305×165	54	11696	233	1.55	223	212	200	176	153	134	118
	46	9899	198	1.52	189	179	168	146	124	107	93
	40	8503	171	1.50	163	153	144	123	103	87	75
305×127	48	9575	195	1.12	165	149	134	110	92	79	69
	42	8196	169	1.09	140	125	111	89	74	62	54
	37	7171	148	1.07	122	108	95	75	61	51	44

Appendix A

Table A2   S275 Steel – 406 deep and below (Continued)

Serial size	Mass (kg/m)	I_y (cm^4)	$M_{c,Rd,y}$ (kNm)	Critical value of L_E (m)	Buckling resistance moment (kNm) for effective length of L_E (m)					
					2	2.5	3	4	5	6	7
305×102	33	6501	132	0.87	96	82	71	54	43	36	31
	28	5366	111	0.85	78	66	55	41	32	27	22
	25	4455	94	0.81	64	52	43	32	24	20	16
254×146	43	6544	156	1.39	146	138	129	113	98	86	76
	37	5537	133	1.36	124	116	108	93	79	68	59
	31	4413	108	1.32	100	93	86	71	59	49	42
254×102	28	4005	97	0.90	73	63	55	43	35	29	25
	25	3415	84	0.87	61	52	45	34	27	23	20
	22	2841	71	0.84	50	42	35	26	21	17	14
203×133	30	2896	86	1.25	79	74	69	59	50	43	38
	25	2340	71	1.22	64	59	55	45	38	32	28
203×102	23	2105	64	0.94	54	48	43	35	29	25	21
178×102	19	1356	47	0.94	39	35	32	25	21	18	15
152×89	16	834	34	0.83	27	24	22	18	15	13	11
127×76	13	473	23	0.74	18	16	15	12	10	9	8

Note
a These sections are part of the Corus Advance range but are not in BS4[46]

Table A3   S355 Steel – 457 deep and above

Serial size	Mass (kg/m)	I_y (cm^4)	$M_{c,Rd,y}$ (kNm)	Critical value of L_E (m)	Buckling resistance moment (kNm) for effective length of L_E (m)					
					4	5	6	7	8	9	10
1016×305	487 [a]	1021884	7775	2.46	6380	5648	5034	4523	4098	3744	3445
	437 [a]	910322	6958	2.41	5630	4951	4381	3909	3520	3198	2929
	393 [a]	807503	6210	2.36	4952	4324	3798	3364	3009	2717	2476
	349 [a]	723131	5725	2.32	4498	3898	3397	2985	2650	2377	2153
	314 [a]	644063	5123	2.29	3976	3425	2964	2587	2282	2035	1834
	272 [a]	553974	4425	2.27	3404	2914	2503	2167	1897	1679	1502
	249 [a]	481192	3916	2.20	2943	2497	2125	1825	1584	1393	1238
	222 [a]	407961	3384	2.12	2474	2078	1751	1490	1283	1120	989
914×419	388	719635	6095	3.38	5855	5474	5095	4722	4362	4024	3712
	343	625780	5340	3.32	5106	4761	4415	4071	3740	3428	3142

Appendix A

Table A3 S355 Steel – 457 deep and above (Continued)

Serial size	Mass (kg/m)	I_y (cm⁴)	$M_{c,Rd,y}$ (kNm)	Critical value of L_E (m)	Buckling resistance moment (kNm) for effective length of L_E (m)					
					4	5	6	7	8	9	10
914×305	289	504187	4337	2.35	3676	3277	2898	2556	2262	2015	1810
	253	436305	3775	2.31	3170	2808	2462	2152	1888	1668	1486
	224	376414	3290	2.26	2729	2399	2085	1806	1570	1375	1217
	201	325254	2881	2.21	2355	2054	1768	1517	1308	1137	1000
838×292	227	339704	3158	2.25	2619	2310	2019	1762	1544	1365	1218
	194	279175	2636	2.18	2146	1873	1616	1392	1205	1054	931
	176	246021	2349	2.14	1888	1635	1399	1195	1027	892	784
762×267	197	239957	2473	2.05	1961	1704	1472	1276	1115	986	881
	173	205282	2138	2.01	1667	1434	1225	1050	908	796	705
	147	168502	1779	1.95	1357	1152	971	821	702	609	535
	134	150692	1649	1.90	1232	1035	863	723	614	529	462
686×254	170	170326	1943	1.98	1513	1309	1128	977	854	756	677
	152	150355	1725	1.95	1326	1137	971	833	723	635	565
	140	136267	1573	1.92	1195	1018	863	735	634	554	490
	125	117992	1378	1.88	1028	866	726	613	524	454	399
610×305	238	209471	2583	2.56	2289	2093	1908	1738	1586	1453	1337
	179	153024	1914	2.48	1664	1501	1345	1202	1076	967	876
	149	125876	1585	2.44	1364	1221	1083	957	847	753	675
610×229	140	111777	1429	1.79	1052	896	764	659	575	510	457
	125	98610	1268	1.77	919	774	654	558	483	425	379
	113	87318	1132	1.74	806	673	562	476	409	357	316
	101	75780	1023	1.68	704	579	477	399	339	294	259
610×178	100 a	72528	961	1.31	470	373	305	258	223	196	176
	92 a	64577	891	1.26	413	323	262	219	188	165	147
	82 a	55869	779	1.23	346	268	215	179	152	133	117
533×312	273 a	198971	2714	2.76	2547	2414	2282	2152	2027	1907	1794
	219 a	151281	2110	2.65	1954	1835	1713	1592	1475	1365	1266
	182 a	123474	1737	2.59	1595	1487	1374	1261	1153	1053	964
	150 a	100840	1431	2.54	1303	1207	1105	1001	903	814	737
533×210	138 a	86088	1246	0.52	898	770	664	580	514	460	417
	122	76043	1103	0.46	780	661	564	488	429	382	344
	109	66822	976	0.41	676	566	478	409	356	315	282
	101	61519	901	1.63	617	513	430	366	317	279	249
	92	55227	838	1.58	556	456	377	318	273	239	212
	82	47539	731	1.55	472	382	313	261	222	193	170

Appendix A

Table A3   S355 Steel – 457 deep and above (Continued)

Serial size	Mass (kg/m)	I_y (cm⁴)	$M_{c,Rd,y}$ (kNm)	Critical value of L_E (m)	Buckling resistance moment (kNm) for effective length of L_E (m)					
					4	5	6	7	8	9	10
533×165	85 a	48505	726	1.25	341	272	224	190	165	146	132
	74 a	41058	642	1.19	279	218	178	149	129	113	101
	66 a	35044	554	1.16	230	177	143	119	101	89	79
457×191	161 a	79779	1304	1.67	986	877	786	709	645	591	545
	133 a	63841	1059	1.61	765	667	586	521	468	425	389
	106 a	48873	824	1.55	564	479	411	358	317	284	258
	98	45727	770	1.54	521	439	374	325	286	256	231
	89	41015	695	1.52	461	384	325	279	244	217	196
	82	37051	650	1.47	415	341	285	243	211	186	167
	74	33319	587	1.46	368	300	248	209	181	159	142
	67	29380	522	1.44	319	257	210	176	151	132	116
457×152	82	36589	625	1.21	348	284	238	205	180	160	144
	74	32674	561	1.19	301	243	202	172	150	133	118
	67	28927	516	1.16	262	207	170	144	125	108	95
	60	25500	457	1.14	223	174	142	119	101	87	76
	52	21369	389	1.11	179	137	110	91	76	65	56

Note
a These sections are part of the Corus Advance range but are not in BS4[46]

Table A4   S355 Steel – 406 deep and below

Serial size	Mass (kg/m)	I_y (cm⁴)	$M_{c,Rd,y}$ (kNm)	Critical value of L_E (m)	Buckling resistance moment (kNm) for effective length of L_E (m)					
					2	2.5	3	4	5	6	7
406×178	85 a	31703	598	1.46	552	510	469	393	331	283	246
	74	27310	533	1.41	485	445	405	333	275	231	199
	67	24331	478	1.39	433	395	358	291	237	197	168
	60	21596	426	1.38	384	350	316	254	204	168	142
	54	18722	374	1.35	334	303	272	215	171	139	116
406×140	53 a	18283	366	1.08	298	261	227	172	136	111	94
	46	15685	315	1.06	254	220	190	141	110	89	74
	39	12508	257	1.02	201	172	146	106	80	64	51

Appendix A

Table A4   S355 Steel – 406 deep and below (Continued)

Serial size	Mass (kg/m)	I_y (cm⁴)	$M_{c,Rd,y}$ (kNm)	Critical value of L_E (m)	Buckling resistance moment (kNm) for effective length of L_E (m)						
					2	2.5	3	4	5	6	7
356×171	67	19463	430	1.38	390	358	327	271	226	192	166
	57	16038	359	1.35	322	294	266	216	176	147	126
	51	14136	318	1.33	284	258	232	185	150	124	105
	45	12066	275	1.31	243	219	196	154	122	100	83
356×127	39	10172	234	0.94	176	150	127	94	74	60	50
	33	8249	193	0.91	141	118	99	71	55	44	35
305×165	54	11696	300	1.35	279	261	242	205	172	146	126
	46	9899	256	1.33	236	220	203	168	138	115	99
	40	8503	221	1.31	203	188	173	141	114	93	78
305×127	48	9575	252	0.97	198	174	153	121	99	84	73
	42	8196	218	0.96	168	146	127	98	79	66	57
	37	7171	192	0.94	145	125	107	82	65	54	46
305×102	33	6501	171	0.76	112	93	77	57	45	37	31
	28	5366	143	0.74	90	73	60	43	34	27	22
	25	4455	121	0.71	73	58	47	33	25	20	16
254×146	43	6544	201	1.21	182	169	155	130	109	93	81
	37	5537	172	1.19	154	142	129	105	87	73	62
	31	4413	140	1.16	123	112	101	80	64	52	43
254×102	28	4005	125	0.78	85	71	61	46	37	31	26
	25	3415	108	0.76	71	59	49	36	29	24	20
	22	2841	92	0.74	58	47	39	28	22	17	14
203×133	30	2896	112	1.09	98	89	81	66	55	47	40
	25	2340	92	1.07	79	71	64	50	41	34	28
203×102	23	2105	83	0.82	64	56	49	38	31	25	21
178×102	19	1356	61	0.82	47	41	36	28	22	18	15
152×89	16	834	44	0.73	32	28	24	19	16	13	11
127×76	13	473	30	0.64	21	19	16	13	11	9	8

Note
a These sections are part of the Corus Advance range but are not in BS4[46]

Appendix B Resistances of UC sections

Compression resistance $N_{b,Rd}$, buckling resistance moment $M_{b,Rd}$ and minor axis resistance of UC columns $W_z f_y$ (S355 steel) are given in the Table B1 below.

Table B1   S355 Steel

Section		\multicolumn{5}{c|}{Storey height (m)}				
		0	3	4	5	6
356×406×634 $W_z f_y$=2310	$N_{b,Rd}$	26260	24365	22784	21100	19305
	$M_{b,Rd}$	4626	4626	4626	4626	4607
356×406×551 $W_z f_y$=1969	$N_{b,Rd}$	22815	21132	19743	18262	16685
	$M_{b,Rd}$	3925	3925	3925	3925	3877
356×406×467 $W_z f_y$=1686	$N_{b,Rd}$	19933	18340	17077	15724	14285
	$M_{b,Rd}$	3351	3351	3351	3332	3265
356×406×393 $W_z f_y$=1391	$N_{b,Rd}$	16784	15384	14295	13128	11887
	$M_{b,Rd}$	2755	2755	2755	2712	2650
356×406×340 $W_z f_y$=1187	$N_{b,Rd}$	14506	13269	12317	11296	10210
	$M_{b,Rd}$	2345	2345	2345	2290	2232
356×406×287 $W_z f_y$=1017	$N_{b,Rd}$	12627	11491	10637	9719	8745
	$M_{b,Rd}$	2005	2005	1996	1935	1878
356×406×235 $W_z f_y$=822	$N_{b,Rd}$	10316	9368	8661	7902	7096
	$M_{b,Rd}$	1617	1617	1598	1543	1490
356×368×202 $W_z f_y$=662	$N_{b,Rd}$	8867	7941	7286	6579	5836
	$M_{b,Rd}$	1370	1370	1339	1287	1237
356×368×177 $W_z f_y$=576	$N_{b,Rd}$	7797	6973	6392	5765	5107
	$M_{b,Rd}$	1192	1192	1160	1111	1064
356×368×153 $W_z f_y$=495	$N_{b,Rd}$	6728	6009	5504	4959	4388
	$M_{b,Rd}$	1023	1023	990	946	902
356×368×129 $W_z f_y$=274	$N_{b,Rd}$	5658	5046	4618	4156	3672
	$M_{b,Rd}$	781	781	760	726	691
305×305×283 $W_z f_y$=785	$N_{b,Rd}$	12060	10425	9365	8223	7069
	$M_{b,Rd}$	1710	1710	1684	1637	1594

Appendix B

Section		0	3	4	5	6
				Storey height (m)		
305×305×240 $W_z f_y$=673	$N_{b,Rd}$	10557	9046	8082	7047	6013
	$M_{b,Rd}$	1465	1465	1425	1379	1337
305×305×198 $W_z f_y$=545	$N_{b,Rd}$	8694	7419	6610	5744	4884
	$M_{b,Rd}$	1187	1187	1141	1099	1058
305×305×158 $W_z f_y$=424	$N_{b,Rd}$	6935	5885	5224	4520	3826
	$M_{b,Rd}$	925	918	877	838	800
305×305×137 $W_z f_y$=363	$N_{b,Rd}$	6003	5080	4501	3885	3281
	$M_{b,Rd}$	792	783	745	709	672
305×305×118 $W_z f_y$=309	$N_{b,Rd}$	5175	4368	3864	3328	2805
	$M_{b,Rd}$	675	665	630	596	562
305×305×97 $W_z f_y$=170	$N_{b,Rd}$	4367	3655	3216	2752	2305
	$M_{b,Rd}$	513	506	479	451	423
254×254×167 $W_z f_y$=392	$N_{b,Rd}$	7349	5908	5060	4197	3418
	$M_{b,Rd}$	836	826	794	765	738
254×254×132 $W_z f_y$=303	$N_{b,Rd}$	5796	4625	3942	3251	2636
	$M_{b,Rd}$	645	629	601	574	549
254×254×107 $W_z f_y$=240	$N_{b,Rd}$	4692	3720	3157	2591	2093
	$M_{b,Rd}$	512	495	468	443	419
254×254×89 $W_z f_y$=198	$N_{b,Rd}$	3899	3083	2611	2139	1725
	$M_{b,Rd}$	422	405	381	357	334
254×254×73 $W_z f_y$=165	$N_{b,Rd}$	3305	2584	2172	1766	1415
	$M_{b,Rd}$	352	334	311	288	265
203×203×127 [a] $W_z f_y$=243	$N_{b,Rd}$	5589	4037	3222	2498	1936
	$M_{b,Rd}$	523	508	488	469	452
203×203×113 [a] $W_z f_y$=213	$N_{b,Rd}$	5003	3594	2858	2209	1710
	$M_{b,Rd}$	459	441	422	405	388
203×203×100 [a] $W_z f_y$=184	$N_{b,Rd}$	4382	3126	2476	1907	1473
	$M_{b,Rd}$	396	378	359	343	326
203×203×86 $W_z f_y$=157	$N_{b,Rd}$	3795	2692	2124	1632	1258
	$M_{b,Rd}$	337	318	301	284	269
203×203×71 $W_z f_y$=129	$N_{b,Rd}$	3119	2201	1732	1328	1022
	$M_{b,Rd}$	276	256	240	224	209

Appendix B

Section		\multicolumn{5}{c}{Storey height (m)}
		0	3	4	5	6
203×203×60	$N_{b,Rd}$	2712	1872	1454	1103	844
$W_z f_y = 108$	$M_{b,Rd}$	233	213	196	181	166
203×203×52	$N_{b,Rd}$	2354	1621	1256	953	728
$W_z f_y = 94$	$M_{b,Rd}$	201	182	167	152	137
203×203×46	$N_{b,Rd}$	2084	1425	1101	832	635
$W_z f_y = 82$	$M_{b,Rd}$	177	158	143	129	115
152×152×51 [a]	$N_{b,Rd}$	2315	1261	879	625	461
$W_z f_y = 71$	$M_{b,Rd}$	156	138	128	119	110
152×152×44 [a]	$N_{b,Rd}$	1992	1073	745	529	390
$W_z f_y = 60$	$M_{b,Rd}$	132	115	105	96	88
152×152×37	$N_{b,Rd}$	1672	889	614	435	320
$W_z f_y = 50$	$M_{b,Rd}$	110	93	84	75	68
152×152×30	$N_{b,Rd}$	1360	714	492	348	256
$W_z f_y = 40$	$M_{b,Rd}$	88	72	64	56	49
152×152×23	$N_{b,Rd}$	1037	524	356	250	184
$W_z f_y = 19$	$M_{b,Rd}$	58	47	41	35	30

Note
a These sections are part of the Corus Advance range but are not in BS4[46]